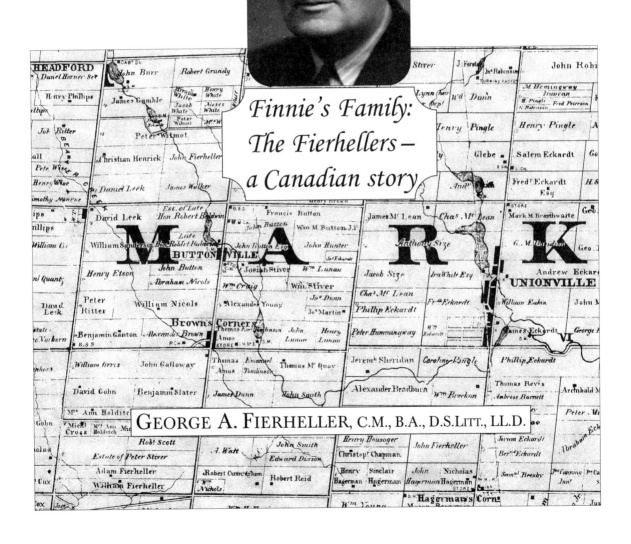

Finnie's Family: The Fierhellers — a Canadian story

GEORGE A. FIERHELLER, C.M., B.A., D.S.Litt., LL.D.

© 2005 George Fierheller.

All rights reserved. Permission to reproduce
in any form must first be secured from
George Fierheller through the publisher.

Library and Archives Canada Cataloguing in Publication

Fierheller, George A., 1933-

Finnie's family: the Fierhellers, a Canadian story / George A. Fierheller.

ISBN 1-894183-69-X

1. Fierheller family. 2. Fierheller, George A., 1933- --Family.

3. Fierheller, George A., 1933-. 4. Telecommunication--Canada--Biography.

5. Computer industry--Canada--Biography. I. Title.

CS90.F5272 2005　　　　　　929'.2'0971　　　　　　C2005-905902-8

Produced and printed by
Stewart Publishing & Printing
Markham, Ontario, Canada L3P 2X3
Tel: 905-294-4389   Fax: 905-294-8718
robert@stewartbooks.com   www.stewartbooks.com

Contents

Acknowledgements 5

Preface 6

I.   The Way We Were 7

II. The Way We Are 29

III. The Way We Might Be 93

The Vierhellers in Hesse 96

A Family Philosophy..97

Further Reading 98

Snippets from the life of George 100

Dedicated to

all the Pioneers

who built this country

Acknowledgements

My thanks to many who provided input or help in assembling this material.

Very useful input was provided by Alf Schenfield, former Deputy Treasurer, Town of Markham & Historian; Markham Union Masonic Lodge; Norman Henrichs, Ancaster, ON; Mark Haacke, Cleveland, Ohio; Thomas Hunter, Navan, ON; 'Bub' Foote, Waterloo, ON; Lorne Smith, Official Historian for Town of Markham, ON; Margaret Godinez (nee Fierheller), Texas; Irene Fowle, North Vancouver, B.C.; and Frank Johnson, Stouffville, ON.

In particular, I would like to thank Ruth Burkholder, Genealogist, Stouffville, ON. and Uwe Porten, Hermeskeil, Germany, for their very professional input.

My thanks to Wendy McKee-Jackson, my Assistant, for typing this rambling manuscript.

And most of all thanks to my wife who tolerated my hours of research.

As you can tell from the many references, Glenna has always been a wonderful stabilizing influence in my life (which I badly needed). I once referred to her as my trophy wife – she deserves a trophy for having put up with me for nearly fifty years.

Preface

"People will not look forward to posterity who never look backward to their ancestors" Edmund Burke.

As one wit once noted "you may not know who your ancestors were but we have been descending from them for centuries".

Another observed "most families are better dead than alive!"

There seems to be a revival of interest in genealogy. I suppose that people are curious about why they are the way they are. Whether we are predominantly the product of our genes or our upbringing or both, it was our ancestors who set the path on which we move.

If this is so, then it may be helpful to know who they were and what they were like.

Most genealogical charts can give you the dry facts of where people were born, when they married, what land they owned and where and when they died. It is more challenging to find out what they were really like. For this reason, this short overview of one branch of the Fierheller family is largely anecdotal. As I am more familiar with what I have done, the latter part is really autobiographical but this helps to link who I am with where I came from.

So that people do not have too high an expectation about what my brief research has turned up, I could give you the conclusion up front.

In everything I have researched, I have never heard a bad word said about the family. But then history says very little about the Fierhellers which in itself is a telling observation. Also people are usually kind about those who have passed on.

One word of mouth observation about the family came from the custodian of the Lutheran cemetery in Buttonville, Ontario. My father had taken me to see this small cemetery where many of the Fierheller family are buried. The elderly gentleman who opened the gate for us and who did not know our family connection commented that "things have been pretty quiet around here since them Fierhellers left".

This ties in with what I have found – a pioneering family that seems to have been honest, hard working, known for having a sense of humour and being generous to a fault. The family members rarely rose to fame and seldom to fortune.

The best I could do personally was to garner a footnote on page 55 of Peter Newman's "The Canadian Establishment", Vol. Two.

In a word the Fierhellers seem to be typically Canadian.

I. The Way We Were

What's in a Name?

No one really knows where my father's nickname, Finnie, came from. My sister, Betty, said that Lady Eaton had come up with it. The Eatons were patients of my grandfather as we shall see. My father's given names were Harold Parsons but the nickname is all that was ever used.

This is the story of his family.

He himself was never interested in his own history. He used to say that the family Coat-of-Arms was a 'pile of manure with two crossed pitch forks'.

Observations such as this and uncountable other jokes came to be referred to as Finnie's Funnies.

But if we cannot trace his nickname, what does the surname mean?

The origin is almost certainly German. I was once told that surnames ending in 'er' were usually from middle-Germany and this is where my search began.

The name was originally spelled Vierheller although it was pronounced as though it started with an F. Early land maps of the Markham area in Ontario showed the original spelling but this was altered over 200 years ago to retain the phonetic pronunciation.

Vier means four in German and a Heller is a small Austrian coin. Perhaps we were usurers in the Middle Ages.

More likely it just meant that we were poor farmers.

Vier can also mean proud. I took some liberties with the English translation of the family name and called our family company Four Halls Inc. but I realize that this is stretching it! The German word for hall is halle.

Some family names give an indication of the early occupation of an ancestor. Examples of this would be Fletcher or Parsons as we shall see. However, as our name does not give any guidance, I settled instead on trying to find where the family had originated in Germany.

Home in Hessen

There are many Vierhellers still in the area near Frankfurt but relatively few elsewhere in Germany. This suggests the family was Hessian.

The Hessians are generally supposed to be descended from the Frankish Tribe of the Chatti whose home was north of the Main River. The Chatti were Christianized by St. Boniface in the early 8th century. But of more interest to the Fierheller history was the introduction of Lutheranism by Philip the Magnanimous in 1526.

This is relevant because the Vierhellers were Lutheran (and as it turned out were also magnanimous in their own way). The topography of Hessen also gives some clues. Even today it consists of richly wooded uplands and small-scale farming is still widespread.

This suggests a farming background for the Vierhellers and this background would have been very helpful when some of the family emigrated to North America in the 18th century.

Waves Across the Waves:

German immigrants to North America came in several waves. However it appears that there were very few Canadians of German-speaking origin who were Reichsdeutsche i.e. Germans coming directly from Germany in the 18th century. According to the Canadian Encyclopedia, most came via other lands.

The motivation for the early wave of German immigrants was the Thirty Year's War (1618-48). However the first recorded German settler in Canada only arrived in 1664.

Most who came to North America settled in what is now the United States.

Between 1749 and 1752, about 2,000 German newcomers landed in Halifax. This was the first organized attempt to settle Germans in Canada.

However, much larger waves of Germans were already flooding into the United States as it would later be known. Early settlers seemed to have favoured New York and North Carolina, the latter being a point to remember as it has a bearing on the Fierheller family.

Settlers before 1730 were usually not Lutheran but were Mennonites, Quakers and Dunkers. This early immigration however amounted to about 13,500 people according to Nelson's book "The Lutherans in North America".

But nearly five times as many German immigrants arrived in North America between 1730 and 1776 with many of these being Lutheran.

William Penn encouraged many to settle in what is now Pennsylvania. This is of course where the term Pennsylvania Deutsch (or Pennsylvania Dutch when the term was corrupted) originated. The Port of Philadelphia Ship Lists show that over 65,000 Germans entered that Port alone from 1727 to 1775.

All this makes it likely although hardly conclusive that the Vierhellers emigrated from Hessen to Pennsylvania during this period. There are still numerous families in the Pittsburgh area using the original spelling.

There is also a group of Vierhellers in St. Louis, Missouri as many of the Germans pushed West. In fact one of the better known Vierhellers was a famous Zoo Keeper in St. Louis in the 1930's.

In 1791, Upper Canada was created (now Ontario) and many American Germans migrated to this new country.

No doubt the offer of free land helped!

A relative with some earlier ancestors!

The Hessians and the American Revolution:

As Ann Galicich noted in "The German Americans" the German mercenaries brought over by the British in an attempt to thwart the American Revolution were even more disliked than the British themselves. There were more than 30,000 German mercenaries who fought with the British during the American Revolutionary War. They are generally referred to as 'The Hessians' although in fact only about 18,000 were actually from Hessen. This raises the interesting possibility that my Vierheller ancestors were part of this group although it is just as possible that they were part of one of the earlier waves of immigrants.

It is possible that the family emigrated as part of this mercenary group. As the patriarch of the Vierhellers in Canada, John Fierheller Sr. (likely Johannes Vierheller) who died in Markham in 1827, would have been of military age in 1776.

However his son, John Fierheller Jr., was born in about 1765 and there is one reference that he was born in the U.S. If this is accurate, then the Vierhellers may have immigrated to the United States in the 1730's - 60's.

There is another even more intriguing possibility. After extensive research in Germany we have discovered a Johann Heinrich Vierheller who was born in Hesse in 1740. What is interesting is that there is no record of his death anywhere in Germany which likely indicates that he emigrated at some point.

Secondly he had a son, Johannes Vierheller who was born in 1764 in Hesse. This date is almost co-incident with the approximate birth date of John Fierheller Jr. who we had supposed might have been born in the United States. There is no record of his death in Germany either leading to the possibility that these were the founding ancestors of our line of Fierhellers who emigrated to North America possibly via Russia around 1770.

The reason for this speculation is that this line of the family seemed to have disappeared entirely from records in Germany. For example there is no record of Johann's son Johannes being confirmed and one would have expected to see this in 1778. It is at least clear that the family left Breungeshain, Hesse sometime between 1768 and 1778.

Although Johann Heinrich Vierheller had three children, none of these appear as a godparent to other relatives in later years, indicating that they had moved far away.

If this is indeed the correct connection then it is possible to trace this line of the family back to a Peter Vierheller who died in Hartmannshain, Hesse some time after 1635 (his wife and three of his children died in 1635 apparently of the plague).

The lineage can be traced quite accurately from Peter to the two Johns that might have been the emigrants to North America. This family chart is on page 96.

The Move to Markham:

Everything to date has been conjecture. We can now move to documented fact.

The Markham area had attracted settlers because of the offer of free land as noted earlier. William Berczy was one of those who took advantage of this. He recruited German settlers and, after a circuitous route that started in Hamburg and wound through Philadelphia and New York, arrived with his group in the Markham area about 1794.

The Berczy settlers included many of the well-known German families that developed the area. The Fierhellers however were not on that list.

John Fierheller Senior (? – 1827):

It appears rather that John Sr. and his family arrived independently as did a number of other Pennsylvania Deutsch families in June 1797. In an Application for Land dated January 1799, he refers to having arrived in the area 18 months earlier. He also refers to himself as being 'German' rather than American. This may have been in answer to a question on the Land Grant Application as to the family origin rather than the location from which they were emigrating.

Little is known of John Sr. other than the fact that he obtained land in the Markham area and died there in 1827. This is recorded in a family bible now in the Markham Historical Museum noting that "John Fierheller from his wife left this world November 27th, 6:00 in the evening one thousand eight hundred and twenty-seven".

John Fierheller Jr. (1765-1850):

John Jr. had apparently immigrated with his father to Markham. In 1801 he married Elizabeth Spring (1777-1861) in Markham. The Spring family was from Rowan County, North Carolina. Her father, Albright Spring, sold his property there in 1796 and received a Land Grant in Markham.

It would appear likely, therefore, that Elizabeth and John Jr. met in Markham and were married there.

Their first child, also named Elizabeth, was born in 1802 – apparently the first Canadian-born Fierheller. This

Grave of John Fierheller, Jr., Buttonville Cemetery, Markham.

younger Elizabeth ultimately married Frederick Peterson whose family started an early Newspaper called "The Markham Advertiser".

John Jr. and Elizabeth had a second offspring in 1803 who was also named John. This seems to be a common practice no doubt to annoy future genealogists.

John Jr. who also applied for and received land in the area took a break from farming to serve in the War of 1812 in the 1st York Militia. Serving with him was a Christian Henricks. John Jr.'s son, John, would marry into the Henricks family in 1828 after first marrying Juliana Peterson (1804-1827) in 1826. Juliana apparently died in childbirth.

Understandably as Markham was a small community, the inter-relationship with the families was inevitable. John Jr. and Elizabeth ultimately had 8 children. John Jr. and Elizabeth are both buried in the St. John's Lutheran Church Cemetery in Buttonville referred to earlier. Many of their children are buried there as well. John Jr.'s gravestone is erroneously carved Feirheller. The spelling with a V had long since been dropped. In fact that spelling was last referenced in the Berczy census of 1803.

John Fierheller (1803-1873):

The first son of John Jr. and Elizabeth was born on 12th of August 1803.

As noted he married Juliana Peterson in 1826. Juliana died giving birth to their only child, yet another John in 1827.

John then married Susannah Henricks (1802-1890) in 1828. Susannah was the daughter of the previously mentioned Christian Henricks Sr. (1766-1840). Her mother was Margarita Fischer (1774-1835).

John and Susanna had 10 children although some did not appear to have survived to adulthood. However, one of their children, William Fierheller, who was born 16th of June 1836 and lived until he was 59, was my paternal great grandfather.

I remember my father, Finnie, telling me that his grandfather, William, spoke no German. Perhaps this is another indication that John Sr. had spent some time in the United States before the family migrated to Markham.

William Fierheller (1836-1895):

William apparently was also a farmer. He married Matilda Haake on 20th of May 1858. Matilda was the daughter of George Haake (1810-1896) and Sophia Quantz (1817-1892). It is likely that one of their sons was named after this George.

George Haake was the son of Johann Haacke who arrived in Niagara-on-the-Lake in 1792. He was only 17 at the time and after working in several jobs on ships arrived in Toronto (then York) in 1800. It appears that he may have worked for William Berczy. He petitioned for land on Yonge Street and when this was refused he applied for and received a grant of land in Markham.

William Fierheller (Finnie's grandfather).

William and Matilda had 4 children. As was typical at the time, a number did not survive their childhood. For example, Joseph Passavant Fierheller lived only from 1870 to 1872. There is a record of an Adaline Fierheller (1859-1887) who married a Thomas H. Munroe but there is no record of any offspring from this marriage.

The remaining two sons George and Clarence Herbert however lived well into the 20th century.

Matilda Fierheller, nee Haake, (Finnie's grandmother).

Uncle Herb (1873-1935) married Laura Heimbecker, another example of the intermarriage of the Germans in Canada. The Heimbeckers were co-founders of one of the largest grain elevator companies with locations across the Prairies. It was called Parish and Heimbecker.

I recall visiting one of the Heimbeckers in the late 1940's in Calgary. Amongst his interests were trying to grow clean potatoes in straw. He had also built an early robot – an odd but charming individual.

I, of course, never knew Uncle Herb but everyone referred to him as an affable, generous individual. As was typical of the German families in Canada, they were noted for their extensive entertaining at the 'groaning' table. He worked for many years for the John Inglis Company.

His wife, Aunt Laura, was a gem. We visited them in Kitchener where they lived at the time on a large property that I recall had deer. Until her death, she always remembered me with a cheque at Christmas.

This then brings me to my paternal grandfather.

'Uncle' Herbert Fierheller (Dr. George's brother).

'Aunt' Laura Fierheller, nee Heimbecker.

<u>Dr. George Fierheller (1860-1942):</u>

The 4th offspring of William was my grandfather who was born in Markham the 30th of August 1860 – a year better remembered for the secession of the Southern States leading to the U.S. Civil War.

He appears to have been the first of the family to break away from the land. He studied medicine at the Trinity College Medical School in Toronto and graduated with a Silver Medal in 1879. He was

Dr. George Fierheller.

Mabel Amelia Fierheller, nee Parsons.

persuaded to take up practice in the Village of Parry Sound in 1884. Parry Sound was then an isolated area with only one other physician, a Dr. Walton. A leading family in Parry Sound was the Beattys. Their patriarch, referred to as the Governor must have been very persuasive to get this young graduate to move to this remote area.

The newly graduated Dr. George Fierheller had just married Mabel Amelia Parsons (1861-1940) on 3rd September 1884. She had been born in North Nibley, England.

We should pause now to look at the Parsons side of the family.

<u>The Parsons & Parry Sound:</u>

While Mabel and George Fierheller were living in Parry Sound, Mabel's sister, Amy Gertrude Parsons (1858-1959) visited them. There she met Walter Reed Foot (1859-1917). Walter was referred to as a 'quiet, reserved and sensitive young man'. He was a chemist and a druggist and built a home on Belvidere Street. Walter himself had been born in Crumlin, County Dublin, Ireland.

Together Walter and Amy had 6 children.

Walter despite being reserved was an entrepreneur. He was the first agent of the new telephone company in town. In 1903 he was appointed Meteorological Observer which required reporting barometric pressure, temperature and other items to Toronto by telegraph three times a day.

Walter and Amy were married on 20th of October 1886. I of course never met him. However, I well remember visiting 'Aunt Amy' in Parry Sound in 1945 and was fascinated to learn that she had carried on as the local weather observer after her husband's death. In fact she became the weather observer for Trans Canada Airlines. Apparently the Foot family made their three observations a day from 1903 to 1945 without a miss for a total of over 38,000 observations. Amy received a special commendation from C. D. Howe for her work particularly during the Second World War.

But where exactly did the Parsons come from? It appears that a John and Marianne Parsons together with their 8 children sailed from England to New York in 1870. I gather that the Parsons despite the name were in the carriage and wagon building business in England. They traveled to Rochester and then on to Cobourg across Lake Ontario. They are buried in St. James Cemetery in Toronto.

John Parsons (Mabel's father).   *Marianne Parsons (Mabel's mother).*

Two of the children we have already referenced – Amy and Dr. George Fierheller's wife Mabel.

A third son was Arthur Rudge Parsons (1850-1941). On 17th October 1880 he joined the Robert Simpson Company in Toronto at a time when the company had only 9 employees. Arthur Parsons was the General Manager and a Director of Simpson's at the time of Robert Simpson's death in 1887. In total he spent 52 years with the Robert Simpson Company.

Another of their children was Florence Parsons. She was unmarried but was referred to as the "business woman of the Parsons family". She also joined the Robert Simpson Company and at the height of her career was in charge of all the cash operations at what by that time was a very large organization.

In his book "A Sense of Urgency", Mr. C. L. Burton who was President of Simpson's Limited writes "the first time I examined the cash office, where money for customers' purchases came through pneumatic tubes, I sat down beside Florence Parsons; and I asked questions she had already many times answered to others. I thanked her and indicated that I would be in again after lunch. She very brightly remarked, "Oh it's alright Mr. Burton, I've seen them all come and go".

Auntie Flo was well respected and loved by her relatives and many friends and was a particular friend of her sister, Mabel. She traveled extensively in Britain, Europe, the United States and Canada.

Another daughter of Uncle Arthur was Vera L. Parsons Q.C. Vera was born in Toronto and attained a Master of Arts Degree in English from the University of Toronto before going to Bryn Mawr College in Pennsylvania for Post Graduate work. She subsequently studied at the University of Rome.

She was a victim of Poliomyelitis and walked with a handicap all her life. She was a wonderful linguist speaking Italian and French fluently.

She then enrolled in Osgoode Hall and graduated in 1924 as a Silver Medalist. She practiced law for the rest of her life. In 1952 a special issue of the Osgoode Hall Law School publication concluded "unquestionably she is the most capable woman barrister practicing at the Ontario Bar".

Back to Dr. George Fierheller:

During his days in Parry Sound, my grandfather apparently visited his patients by horse drawn sled in the winter. He described his frustration at being unable to cure many who died of diphtheria and similar diseases that were untreatable at the time. He also mentioned that he was often paid with chickens (perhaps where the name 'chicken feed' came from!), firewood or whatever his patients could afford.

Small wonder then that after several years he and Mabel moved first to Sunderland, Ontario and then to the Markham area. He set up a practice in Unionville in what is referred to as the Doctor's House that still exists on Main Street. It is now a restaurant. A Doctor Eckardt also practiced there, another well known Markham name.

Dr. George Fierheller was an active Mason. The Markham Union Lodge was the 12th oldest in the Toronto area having been established in 1857. My grandfather served as Master in 1894/5.

They lived at the time in nearby Markham on Main Street. It was in this house that my father was born on the 1st of September 1892.

Dr. George and his family finally moved to Toronto and took up residence at 535 Sherbourne St. My grandfather had a long and successful medical practice serving many of the prominent families in Toronto at the time including Sir Henry Pellatt of Casa Loma fame (I still have an ivory carving given to my grandfather by Sir Henry); the Eatons (hence the story that it was Lady Eaton who gave my father his nickname); and the Masseys (one of whose sons Raymond would later serve with my father in Russia in 1919).

The family moved to Queen's Court Apartments on Jarvis Street which at that time was the most fashionable street in the city. Dr. George was clearly an innovator. One of the products he developed was called Salzo, a purge but also somewhat like an early version of Alka Seltzer.

Another was a chewy chocolate flavoured food supplement called Olajen.

I particularly remember the latter product as it was manufactured by a Toronto confectionery firm called Robertson Brothers Limited on Queen Street East. My grandfather took me on a tour of their factory and I was fascinated to see how they made various kinds of candies. This was good for several 'show and tell' sessions at school. I was truly a kid in the candy shop.

I remember however being somewhat disillusioned to see them making marshmallow Santa Claus's in August!

Despite his entrepreneurial activities, Dr. George made a number of questionable investments (a ranch out west where the cows froze to death in an early snow storm, a citrus plantation in the Caribbean that was wiped out in a hurricane and shares in something called Patricia Airways!). He was a successful and well loved physician but no businessman.

I remember him as the classic example of a true gentleman. He was bright but quiet and had what seemed to be the Fierheller sense of humour and a particular sense of humanity. After his wife died he lived with us at 49 Woodlawn Ave. Even in his final years, he would help the neighbourhood kids build a hockey rink in our backyard.

He apparently was hardly ever sick and I remember him saying that he had trouble understanding what a headache felt like as he had never had one. He died when I was 9 and apparently his last words were "that was funny" to describe the stroke he had just had.

In many ways, he was my first role model and I remember him very fondly.

He and his wife Mabel are buried in Mount Pleasant Cemetery in Toronto.

Harold Parsons (Finnie) Fierheller (1892-1968):

George and Mabel had two sons, Herbert Stanley (1886-1910) and my father.

Uncle Stanley studied Electrical Engineering at the Faculty of Applied Science at the University of Toronto. He was described as being "one of its brightest younger members". He graduated with the highest standing and went on to teach at the University. I have a copy of one of his papers that was published in the Bulletin of the Royal Society of Canada in 1909.

The young Finnie, Markham, 1890s.

Stanley Fierheller (Finnie's brother).

Tragically he contracted tuberculosis and died in the early summer of 1910. I have a letter written to my grandfather by Robert W. Falconer, President of the University of Toronto expressing his regrets at the loss of someone "having such a promising career".

Finnie had his early education in Markham but after moving to Toronto, attended Jarvis Collegiate.

He recounted one of his memories of his early days in Toronto when he was apparently run over by one of the first electric cars in the city. Upon graduation he held various jobs including operating the Parimutual at the old Woodbine Racetrack where he worked with Paul Martin Sr. who had a summer job there.

At the outbreak of World War I he enlisted in the Army Service Corps. Perhaps because of his perceived background in rural Ontario, he first served as a Calvary Officer. He became a riding instructor at Val Cartier, Quebec and his training duties took him literally across the country from Halifax to New Westminster, British Columbia. *(See photo on next page.)* It was while in Halifax that he met his wife to be.

It was customary for the young belles to entertain the troops at Tea Dances. My mother, Ruth Hatheway Bauld (1897-1989) told me that at one of these Tea Dances, she was the partner of the Duke of York, later, in 1936 to become King George the Sixth. She described him as being "an awkward young man who stepped on my feet".

More importantly, it was at one of these where she met my father. She apparently commented after the event that she had met a young Lieutenant with a 'peculiar German name'. Despite this apparent drawback, they were married shortly thereafter at St. Andrews Church, Halifax on December 7th 1916.

Officers and NCO's, New Westminister, B.C. (Finnie front right).

The Bauld Facts:

My mother's family was Hatheway, spelled with an 'e' making the name somewhat rare. The Hatheways, at least according to my mother, provided the aristocratic side of our family. Evidently her mother had accumulated a great deal of the Hatheway family history tracing our ancestors back to Sir Walter Raleigh and Humphrey Gilbert. She had mentioned that one of their ancestors was Dame Sybil Hatheway of Sark.

Hatheway Crest (lion holding a fleur de lis).

Hatheway Shield.

Her mother, Elizabeth, was the daughter of a Dr. Joseph Hatheway whose wife was the daughter of a French Count. My mother recounted that the family had a Coat of Arms with the motto Frangi Non Fractum (Bent But Not Broken). The family evidently originated in England in the Forest of Dean area.

All that dubious history aside, the more easily traceable part of the family was when Ebenezer Hatheway and his family emigrated from the United States as United Empire

Loyalists in 1783. They settled in tents along the Saint John River together with many other Loyalists.

D.G. Bell in a book titled "Early Loyalist Saint John" had the following quote describing the conditions at the time.

"An iron shore, ordained by fate
For Loyalists, their last escape"
A spectator 1784

One arrival called the area that would later become Saint John the "roughest land I ever saw".

Most Loyalists arrived by ship from New York between 1782 and 1783. The Royal Forces quit New York 25th of November 1783 and only a small number filtered into the New Brunswick area after that.

The Hatheways however prospered. One son of Ebenezer was 'a wealthy man, owning a large amount of property in Saint John'.

A Thomas G. Hatheway (1791-1855) married Ann Canby on 30th of April 1815. He was the youngest son of Ebenezer. My mother had a brother Alfred Canby who died in infancy (one year old) and this seems to confirm the family relationship back to Ebenezer.

Many of the Hatheways are buried in the Old Church of England Burying Ground in Fredericton, New Brunswick.

Another of Ebenezer's sons was Calvin Luther Hatheway (1786-1866) who authored a "History of New Brunswick". His son Frederic William Hatheway (1811-1866) carried on a successful business that his father had started in agriculture which included substantial trading in ships that they not only owned but built.

A younger son of Calvin Luther was George Luther Hatheway (1813-1872) who was very active in New Brunswick politics. He was appointed Leader of the Government and Provincial Treasurer in 1871.

According to Isabel Louise Hill in Volume I of the "Old Burying Ground Fredericton, New Brunswick" there are 13 Hatheway gravestones in the old cemetery still in good condition. However, the cemetery record stops before a firm link can be established to my grandmother, Elizabeth.

Elizabeth Hatheway, known as Lizzie, was apparently a vivacious and attractive woman who snared a young member of the Bauld family of Halifax. It appears they were married in Saint John in the late 1880's.

My grandfather, Alfred Mason Bauld (1865-1944) was born in Halifax, the son of William Bauld (1824-1882) and Emily Sawyer Gray (1831-1913). William and Emily married the 7th of June 1856.

It is uncertain as to exactly when the Baulds emigrated to Canada but there are Bauld gravestones in the Old Cemetery at the foot of Spring Garden Road in Halifax. Two of these strangely enough are for a John Bauld and both stones are dated 1826. Two other stones dated 1829 are for a William Bauld and a Rupert Bauld.

My great, great grandfather, also a William Bauld, was born in Dunfermline, Scotland in 1792. It is likely that he immigrated to Canada in the early 1800's and settled in the Halifax area. The other famous resident of Dunfermline was Andrew Carnegie although unfortunately there is little indication that his capitalistic genius rubbed off on the Bauld family.

William Bauld married Elizabeth Trider (1800-1878) on the 4th of August 1820. This suggests that some of those referenced earlier who died in the 1820's were children of this marriage. This William died in 1862.

So the Hatheways appear to have arrived in Canada in 1783 and the Baulds likely around 1815.

Alfred Mason Bauld (1865-1944):

My grandfather was always called AM or simply the Major. The family was very entrepreneurial and had established a prosperous importing firm called Bauld, Gibson & Company with offices on Barrington Street in Halifax. The company had been started by his father, William and his uncle, John Gibson.

The firm imported rum, tobacco and other less corrupting items such as flour, sugar, molasses, China tea and even clay pipes. However, AM's elder brother, Henry G. Bauld, entered the firm in the 1880's as a Junior Partner and as he was the eldest son was slated to take over the management.

Major A.M. Bauld.   *Elizabeth Bauld (nee Hatheway).*

AM, instead of entering the firm enlisted in the North West Mounted Police. The Indians in Western Canada were growing increasingly restive as a result of the disappearance of the buffalo and crop failures in a Saskatchewan valley. The Federal government enlarged the NWMP to 1,000 men to meet this challenge.

My grandfather was sent West and fought in the North-West Rebellion of 1885 when only 20 years old. During the conflict, some 38 Government Troops and 72 Métis and Indians died. The rebellion ended with the hanging of Louie Riel on the 16th of November 1885.

My nephew, David Lindop, still has AM's ceremonial sword from his service in the North West Mounted Police.

AM then returned to Halifax where he met Elizabeth Hatheway.

The newly married couple settled in a home on the Northwest Arm in Halifax. It is likely that AM joined the family firm at that time, possibly as a traveling salesman as the firm had embarked on a major expansion. At one time (according to a 1935 interview with Henry G. Bauld in the Maritime Merchant) the firm had as many as 14 men on the road.

Typical of the business practice of the time, "accounts were settled twice a year". This was a precarious procedure at best although as my Great Uncle Henry reported "no one was ever dunned".

The First World War intervened and AM enlisted. Perhaps given his NWMP experience and his place in the community, he was promoted to Major and served with the 1st Canadian Garrison Artillery. He was responsible for the Halifax Citadel Forces.

By the end of the War, Bauld, Gibson & Company had declined and after 107 years finally folded. There were rumors that Uncle Henry had liberally drawn on the firm's resources while AM and others of the family were serving in the Armed Forces but the truth of this is lost in family lore.

AM was a fine athlete. He excelled in snowshoeing (I still have his silver cups) and rowing. He was the Founding President of the Halifax Amateur Boating Club in 1904 and one of the Founders of the Waegwoltic Club. The Wag still exists. It was one of the first clubs anywhere to grant women members equal rights.

As his middle name implies, the Baulds were Masons. They were also Calvinists. As noted earlier, my paternal grandfather, Dr. George Fierheller was also a Mason. This tradition from both sides of the family did not however pass on to any of their descendants!

After the demise of Bauld, Gibson & Company, AM became a salesman for a major pharmaceutical firm. In 1934 he moved to Toronto where my parents had also settled.

I remember him as a very kind and generous man. On occasion he would walk me to kindergarten at Brown Public School on Avenue Road and I recall he used to read stories to the kindergarten class.

He passed away on the 4th of July 1944 at Christie Street Hospital although he was actually buried in Halifax in Camp Hill Cemetery.

Ruth Hatheway Bauld (1897-1989):

My grandmother, Elizabeth Hatheway, was apparently known as the Belle of Saint John. She had 3 children, Canby, who died as an infant and my mother's sister, Gretchen, who was 4½ years older than my mother. Gretchen apparently painted beautifully and played the piano with great skill. However, she was never too strong. She married a minister and spent many years in Newfoundland. The climate was damp and she died at 36.

My mother followed in her mother's footsteps. She was very outgoing and loved to entertain. Like her father, she was a sports enthusiast who would walk five miles to the Northwest Arm to skate. She played tennis and golf with great enthusiasm.

School however was not her favourite place. She attended the Halifax Ladies College and then moved on to a boarding school at Edgehill. The school was run on strict British lines and she hated it. In fact, she ran away and returned to Halifax.

The family spent their summers at Birch Cove, an inlet off Bedford Basin. They even took along the family piano each year.

The Royal Canadian Navy played a major part in her life. One of the early First World War volunteers was her cousin, Victor Hatheway. However, my mother had many friends among the cadets at the Royal Canadian Naval College H.M.C.S. Niobe. A number went on to become Admirals and Captains.

The 1914-18 War however brought its own tragedies. Four of her close friends including Victor Hatheway had volunteered for sea duty on the British battleship H.M.S. Goodhope. The ship was sunk in the Battle of Coronel off South America. They were the first Canadian naval casualties of the war.

Now it is time to return to Finnie.

Finnie & Ruth:

My father's war service continued after the 1918 Armistice. Canada had gained a considerable international reputation in World War I and it was perhaps to maintain this momentum that Prime Minister Sir Robert Borden agreed to send Canadian troops to Russia to aid in the effort to crush the Bolsheviks.

Finnie was part of that contingent that sailed from Vancouver on the R.M.S. Empress of Japan. I remember him describing to me the very rough Pacific voyage made even more difficult as the horses as well as the men got seasick. They arrived in Vladivostok on October 27th 1918, only a couple of weeks before the war ended in Europe.

Ruth and Finnie, 1920s.

As described by Roy MacLaren in "Canadians in Russia 1918-1919", the contingent in fact accomplished very little. They were only involved in one action that took place about 50 km. north of Vladivostok. There were no Canadian casualties in the single engagement although typically of the times, 3 died in accidents and 16 from disease.

The adventure was not without its rewards for the Fierheller family however. Amongst the articles Finnie brought back were a beautiful set of Japanese ivory carvings and two Russian cavalry swords all of which I still have.

There were many stories of my father's service in the armed forces. My father recounted trying to keep warm in the vicious Russian winter. When a balky stove refused to stay lit, he apparently poured some coal oil into it. The stove blew its lid and blew the newspaper out of the hands of the aforementioned Raymond Massey leaving him holding just to two portions in his hands! Raymond Massey went on to become a famous movie actor having survived this incident.

On his return, my mother told me of another incident when my father, who had been drinking with the other officers, fell into their bed with his spurs on. Their marriage survived that incident as well!

After the War my father who had short employment stints with "every organization on King Street" joined Ontario Hydro. He was asked to establish a new division to handle that huge organization's printing and blueprinting. In those days the outsourcing craze had not yet taken hold. Later when aerial photography became increasingly useful to Hydro, the processing of photos was added.

An early associate of my father's was a painter who did calendar art. His name was Frank Johnston. Frank later changed his name to Franz after becoming a member of the Group of Seven as he thought the name was more appropriate to a painter.

The family was now comprised of two daughters, Betty Joyce (1918 -) and Ruth Audrey (1920 -).

Betty was a talented pianist and a very good student. However, tragedy was to strike in 1930 when she was stricken with encephalitis. She never recovered all her faculties after a 30 day coma with high temperatures. Dr. George Fierheller could only express his helplessness in those days before penicillin. Betty stayed at home after her partial recovery for a number of years but is now institutionalized in Newmarket, Ontario.

Finnie, Betty, Ruth and George, Thunder Bay Beach.

Audrey was a vivacious and attractive young lady who was working at Eaton's when the War broke out. She enlisted as a Wren and was stationed in Halifax. It was there she met a young Lieutenant Commander in the British Navy, John L. Lindop. When John returned to England with the Navy Audrey somehow managed to get a transfer to the British Wrens and got transport on a merchant ship in the middle of the Battle of the Atlantic. They were married in Campbeltown, Scotland.

Auddie's stories of adapting to life in wartime Britain could fill a book. She described her first meeting with her new father-in-law. She arrived at their home in her Wren's uniform and found him in the garden. He looked up and said "you must be Audrey. Well make yourself useful and help me pick these beans for dinner".

Audrey and John with John's brother Jeff and his wife Birdie.

This was the story of a reverse flow war bride. Fortunately the Fierheller sense of humour saw her through this difficult period. However, while John was on active duty in Malta and elsewhere, Audrey finally was transferred to London. London was still under attack by German bombers and she brought back to Canada after the War a piece of shrapnel that had just missed her. The house next to hers at Catford Bridge in South London was demolished by a bomb. However, both survived the War although John lost one of his two brothers in an accident in Ceylon.

John stayed in the Royal Navy after the War but eventually decided to immigrate to Canada. He had attended Rugby School in England and this plus his naval experience gave him the background to apply for and become the Bursar at Trinity College, Port Hope.

John and Audrey had three sons, David, Michael and Peter. David and his wife, Audrey are keen bridge enthusiasts. Audrey has written a series of books under the name Audrey Grant including the *Joy of Bridge*. Michael was head of the English department at a local high school and Peter was Branch Manager for the Toronto-Dominion Bank in Ottawa.

John and Audrey still live in Port Hope.

Finnie and Ruth and the family lived at 49 Woodlawn Avenue West in Toronto from the mid 20's until the end of the Second World War. Like many Canadian families, we took in war guests, Peter and Ivan Foster and their mother.

Woodlawn Avenue was at one time a very fashionable neighbourhood but by the 1930's was in decline. However, because of its central location in Toronto it is now once again a very 'up market' area.

However, not waiting for this turnaround, in 1946 my parents moved to 36 Hudson Drive in Moore Park.

Finnie and Ruth were active golfers playing first at Thornhill Golf & Country Club and later at York Downs. They loved bridge and entertained regularly.

Finnie and some of his friends started a Wednesday night Poker Club that ran for over 30 years – the longest known span for any such club.

Of course, from my standpoint the most significant event of the depression in the 30's was my birth on the 26th of April 1933. This was 15 & 13 years respectively after the birth of my two sisters and I suspect was likely the aftermath of some cocktail party!

The timing turned out to be fortuitous. While most people naturally credit their accomplishments to hard work and great intelligence the authors of Boom, Bust & Echo pointed out that anyone born in the 30's when there were very few children was almost bound to succeed as we were such a scarce commodity!

My parents chose to name me after my two grandfathers, both of whom were still alive at the time.

So we are now up to the 'Contemporary' Fierhellers!

Finnie's retirement party, Ontario Hydro. Tom Allen & wife, Ruth, Finnie, Glenna & George.

II. The Way We Are

<u>George Alfred Fierheller (1933 -):</u>

I was raised almost as an only child and was likely spoiled rotten. I remember being somewhat shy and my parents telling the story of my meeting at an early age with some people they were entertaining. When I was prompted to say 'hello', I replied that "we have not been properly introduced".

As well as being a stuffy little brat I was a hopeless athlete. In fact, I was not only un-athletic but anti-athletic. I struggled through Physical Education but fortunately was an adequate student and managed to get through public school.

A near miss occurred when I was eleven years old. It was approaching Christmas, 1944 and I was feeling increasingly sick. Our family doctor diagnosed this as stomach flu but it turned out I had a badly ruptured appendix. I was rushed to Sick Kids Hospital where the Chief Surgeon a Dr. LeMesurier operated immediately but advised my parents that there was a slim chance that I would survive. Fortunately no one told me and although I had to spend a month in the hospital on what was close to life support, I was obviously not ready to pack it in. Sometimes it is just as well not to know that everyone else has given up hope!

For my last year, in grade 8, I transferred from Brown Public School to Whitney Public because of our move to Moore Park. I had applied for entrance into the University of Toronto Schools and after passing the entrance exams was accepted. This meant I did not have to write my final Matriculation exams at Whitney in 1946. I was the envy of the class being able to leave for summer holidays early.

UTS was then and still is an institution with rigorously high academic standards. We were generally considered to be somewhat elitist or for those less kindly disposed, just plain 'nerds'.

I became the Literary Editor for the Year Book, the "Twig", and avoided some Phys Ed by joining the rifle team. This was the only sport I had found that you could do lying down (well almost the only sport!). It turned out that I was quite a good marksman and attained a Silver Dominion Marksman's Medal. Otherwise my years at UTS were reasonably undistinguished. I did however come in second in an extemporaneous speaking contest which perhaps foreshadowed some of my later activities.

In the graduation Year Book, I stated my ambition was "to become a millionaire". As most of my UTS fellow students were planning to be doctors, lawyers or university professors, this was felt to be a trifle materialistic!

FIERHELLER, GEORGE — George is interested in tennis, golf, and record collecting. Intends to become a millionaire via C. & F. or Pol. Science. His love life and summer are a complete mystery.

George states his ambitions in the U.T.S. "Twig" (Year Book).

Despite this stated ambition, I applied to and was accepted at Trinity College at the University of Toronto. My father encouraged me to take something practical like Commerce & Finance. I wanted to take Philosophy & History. We compromised on Political Science & Economics – neither of us having any clear idea of what Economics really was. Four years later, I still did not have much of an idea but had a wonderful time at the College. Trinity at that time still required the men of College to take Religious Knowledge, the only effect of which was to encourage me to become an Agnostic. At Trinity, undergraduates were still required to wear gowns and actually did have occasional afternoon teas with the Provost.

I was rushed for and joined Sigma Chi Fraternity. This turned out to be a wonderful opportunity for lasting friendships. One of these was with Ted Rogers who was also a student at Trinity.

Ted and I became involved in the Mock Parliament at the U of T. Ted was always actively involved in Conservative politics. I remember he and I were hammering up election signs where these had been requested by Conservative supporters. Ted took the opportunity to place the signs on most of the houses on the block whether requested or not. When I queried this approach, he just noted that if they were Conservatives then we had done the right thing and if not they would take the signs down so no harm done. This was a forerunner of the usual Rogers aggressive approach to business and everything else!

Like most students, I had to earn part of my tuition and of course took summer jobs. The first was as an Income Tax Assessor working at the National Revenue building on Front Street in Toronto. I remember being given a pile of tax returns to assess and by about 10:30 in the morning had completed these and asked my Supervisor what I should then do. His advice was 'slow down'. This was my first exposure to union working rules.

I therefore spent much of the summer reading books on topics that had nothing to do with my course, e.g. Glasstone's Source Book on Atomic Energy and similar light tomes.

Despite taking a Liberal Arts course, I maintained a strong interest in Science which was to stand me in good stead later on.

Another summer job involved being part of a survey team in Northern Ontario relocating power lines north of Lake Superior. Our Camp Boss was a crusty old northerner who thought he should make some effort to educate the city-types assigned to the survey team. While walking by the lake, he picked up a piece of wood intent on teaching me to recognize various trees. He asked what type of wood it was. I helpfully replied 'driftwood'. He looked at me in disgust, threw the piece of wood back in the lake and that was the end of my nature training. However, I returned to the survey team the following summer. It was actually a great experience seeing another part of Canada. I was stationed for a while in Sudbury where we stayed at the old Nickel Range Hotel, long since torn down. This experience gave me a real appreciation of the damage people could do to the environment. The landscape resembled a desert in those days as a result of the fumes from the smelters.

The Grand Tour:

The aforementioned Auntie Flo had died leaving me $500. I felt that I should round out my education by using this money to tour Europe. I mentioned to my father that I had always wanted to see Europe to which he replied that he had never been and in my case 'always' was a short time as I was just 20 years of age. Besides he correctly pointed out that $500 would not take me too far even in those days.

I reasoned however that it would pay for a trip across the Atlantic in steerage; I could stay with my sister, Audrey, who was still living in London and if I stayed in youth hostels and at universities, could somehow manage with the bit that I had saved from my summer jobs.

With some help from my parents, I booked passage on a small French liner called Le Flandre. It sailed from New York to Portsmouth.

I shared an inside cabin near the engines with several other young students, mostly from the U.S. We ate some meals with the second class passengers and all of us greatly admired a vivacious young dark haired woman who was forever invited to the Purser's Table. She was very eye catching with large hoop earrings and what we all acknowledged was a fabulous figure. In fact we concluded amongst ourselves that we would all like to pursue her but of course that was out of the question for impoverished students. More on this is to come.

When I disembarked at Portsmouth, I was met by my brother-in-law, John who took me on a pub crawl and subsequently entertained me with his fellow Officers in the Mess. They laughed at this landlubber from Central Canada when I remarked that the ship was much higher when I got off than when I had got on. Naturally I had forgotten about tides.

I then went to London, a city I immediately fell in love with and to which I return regularly.

However my ambition was to see the Continent. I booked a fourth class railway tour ticket and embarked for Paris.

Like any tourist, one of my early visits was to the Louvre. By coincidence I recognized the above mentioned and much admired fellow passenger from Le Flandre. I introduced myself and we soon discovered we shared an interest in the Arts, Music and History. Her name was Bronia Oblatt. Despite this Catholic sounding name she was actually a Polish Jewess although she was married to a Doctor in Mexico City. She was traveling on her own as her husband had little interest in the Arts or travel.

We decided to go together to the Musée Rodin and other locales in the next couple of days. Although she was dramatically better off financially than I, she agreed to share my relatively humble lifestyle. This included a lunch of wine, bread and cheese in a Paris park.

We were enjoying our lunch when two men in plain clothes approached us. They shouted something at us in French and one grabbed her by the arm.

Having seen any number of John Wayne movies, I did the chivalrous thing, pulled him away from Bronia and for the first and only time in my life hit the man on the chin. I am not sure how John Wayne managed his barroom brawls but clearly I did not have the technique well developed. The man fell backwards obviously not hurt while I nursed my aching hand. I need not have bothered.

Bronia was quite capable of looking after herself. She wore a large square cut ring on her right hand and took on the other individual. One backhand slap left him with a nasty cut across the face. By this time the first man was pulling out some identification and we discovered that we were dealing with the Sureté. It turned out that we were sitting on the grass which was apparently not allowed. The result was that we were arrested for resisting arrest, assaulting a police officer and anything else they could think of. The two of us were bundled off in a police vehicle complete with wailing sirens. At the station we waved our passports, demanded to see our Consul and generally made as much of a nuisance of ourselves as we could. Finally, the police let us go on the understanding that we would be out of France within 48 hours. Despite the fact that this cut short my visit to Paris, Bronia and I had 'bonded' and she agreed to also get a fourth class rail ticket and travel with me to see the rest of Europe.

The next evening we boarded the overnight train to Milan. The fourth class compartments were designed to seat 10 people on wooden benches and were anything but comfortable. The two of us decided to seek better accommodations and walked up to the first class carriage. When we awoke the next morning two uniformed men were staring at us. One of them shouted Achtung! I commented to Bronia that I did not speak Italian but this did not sound right. It turned out that in the middle of the night the fourth class carriage had indeed carried on to Milan but the first class carriage was routed to Bern, Switzerland. They threw us off the train and we spent a delightful few days touring Switzerland in Interlaken and other locations.

We finally did get to Italy and spent several weeks in Rome, Florence and Venice.

I had obviously discovered a method of 'touring Europe on $5.00 a day' as Bronia picked up most of the charges for hotels, meals, etc. This was an approach not even Thomas Cook had thought of!

We parted in Rome and she flew back to Mexico City. I was now back to traveling fourth class and after several other adventures arrived in Madrid. I had made arrangements here to meet a friend from Toronto, Greta van Valkenberg. Greta's mother ran a well known dance school to try to educate private school boys and girls how not to step on each others' feet while dancing at school proms. Greta was a very attractive brunette but my hopes were dashed when on arriving at the agreed-to hotel, she had left me a note indicating that she had eloped with the Norwegian Vice Consul to the Canary Islands. This was hardly a major international diplomatic posting but was clearly a step up from a penniless student.

I wended by way through Spain and back through France noticing that I was losing considerable weight. When I finally arrived back at my sister's home she took one look at me and called the doctor. It turned out that I had contracted dysentery. On hearing some of my stories, my ever observant sister noted that "you were lucky that was all you caught".

The doctor correctly concluded that I had contracted dysentery and ordered me to an isolation hospital in London. He also required my sister to fumigate the house. It was hardly a welcome return by her errant brother.

In the isolation hospital, we were in a ward with others having polio, meningitis or whatever else people had contracted. The hospital was clearly not going to waste their still short supply of penicillin and put me on sulfa drugs. This required my taking eight pills every six hours. I was sure I was going to rattle.

One of the ingestion of pills was at 6:00 a.m. There was a very cute young Irish nurse who attended me. At one 6:00 a.m. session she woke me by pouring part of a pitcher of water on me. I responded to this indignity by picking up a towel, dipping in the end of the rest of the water and chasing her down the hall. "Die earthling" I shouted as I flicked the towel at her. As she rounded a corner she slipped and ended up on her posterior with her nurse's cap askew. To our horror immediately in front of us was the Head Nurse, the Head Doctor and an entourage of Residents doing the morning rounds. The Head Nurse took one look at me in my open at the back hospital gown and in her usual stern British manner simply said "Young man you get back to your room. Young lady you come with me!" I never saw her again. I presume they shot her!

My grand tour continued once I was discharged and I managed to extend what had developed into a major learning experience as long as I could. I finally caught the return voyage in the middle of November and returned to Trinity. I had missed the first part of the academic year. Despite spending time on the Hart House Art Committee and other activities, I managed to laze my way through my final year and graduated with an Honours Degree in Political Science & Economics.

George in the Job Market:

I have always been influenced by the women in my life – wives, mothers, daughters, etc. As I came close to graduation, my mother gently reminded me that perhaps it was time to start looking for a job. I had studiously avoided all the 'on campus' interviews, not wanting to rush my choice of a career. However, by March of 1955 I was reading Time magazine in the Junior Commons Room at Trinity and ran across an article called Klink, Klank, Think. It was the story of the Watson family and IBM. The description of Thomas Watson Sr. with his THINK signs and the incredible machines he had developed intrigued me.

I wandered down to 36 King Street East which was the IBM Sales Office in Toronto and suggested they needed to hire someone like me. After passing the Programmers Aptitude Test and being scrutinized in about half a dozen interviews up to and including the Vice President of IBM Canada, I was offered a position as a Sales Trainee. I had applied for a couple of other jobs including being a trainee to become a Chartered Accountant with what was then Price Waterhouse. However, IBM offered a starting salary at an astounding $275 per month and they won the day.

I often remarked later that if I had gone one storefront past the IBM office, I might today be the head waiter at Letros Greek restaurant.

This was the start of an interesting if totally unplanned career.

It was a long way from my other ambition which was to 'reform the world'. However the only thing my Political Science Degree seemed to equip me for was to be Prime Minister and there already was one of those. Having settled on this somewhat commercial start to my career, I wrote to Bronia with whom I still corresponded that I was already "a 21 year old failure".

It turned out that I had made a very fortuitous choice. While the backbone of IBM's business at the time was in punched card or, as it is sometimes referred to, Unit Record equipment as well as typewriters, time clocks and other business equipment, the computer age was already on the horizon.

My first step however was to complete a rigorous training course on how to apply machines to business problems. This was held at Endicott, New York to be followed after some field experience by sales school held in Poughkeepsie, New York. The total training lasted nearly two years. In good IBM tradition we started each school day singing company songs such as *Ever Onward* and *Hail to the IBM*. Pictures of Thomas J. Watson Sr. hung everywhere like Big Brother. However the rigor and high ethical standards that we learned in those training years shaped my whole view as to how one should conduct business. Amongst the many lessons was that if you undertook a commitment, you met it. More than any electro mechanical or electronic developments that IBM undertook, it was this sense that a customer could rely totally on the company that was the backbone to its success.

Upon returning to Toronto I was assigned a Sales Territory working with the then ubiquitous Unit Record Systems.

We should now pause however for the next major event in my life.

Glenna Elaine Fletcher (1933 -):

During my latter years at UTS and my early days at Trinity, I had "gone steady" with a lovely young lady named Liz Holwell. Liz had attended St. Clement's school but had a good friend who was attending Branksome, Glenna Fletcher. At that time Glenna was dating a UTS friend of mine, Russ Howland.

After my European adventure, Liz and I drifted apart. Glenna and Russ did the same. Glenna and I started dating in the mid 50's and once again my mother gently prodded me into action. "This one is too nice to let get away" she observed.

I agreed and the wedding was set for 17th of April 1957. It was referred to by Birks, Eaton's and other wedding registry stores as the Fletcher/F wedding as no one would risk pronouncing Fierheller.

The ceremony was held at Kingsway Lambton United Church. This was a concession to our parents because as it turned out Glenna shared my lack of reverence for any organized religion and this was practically the last time we were ever in a church. The reception was held at the Old Club House at Lambton Golf Club.

Liz was one of Glenna's bridesmaids and Russ was one of my groomsmen. Somehow we managed to get through the ceremony by saying yes to the right people at the right time! Glenna had had an eclectic early career. She had been a fashion model and then had tried to follow her mother into nursing. However she found that teaching was what she really enjoyed. She attended the Toronto Teachers College and at the time of our marriage was teaching at Keele Street Public School.

Glenna and George Fierheller's wedding party, 17th of April, 1957.

*Back row: Bob Fielden, Grant Fletcher, Judy Fletcher, Joan Johnston, Glenna & George, Bill Corcoran, Ted Stephenson, Hugh MacKenzie, Russ Howland.
Front row: Carol Perry, Liz Holwell.*

Glenna had one brother, Grant, who subsequently married Judy Jenkins, daughter of Ross Jenkins, a Sr. Vice President at Eaton's. Grant had gone through Upper Canada College and subsequently worked at Eaton's in various managerial positions.

Glenna's Family:

Glenna's parents lived in the west end of Toronto, first on Glendonwyne and then at 21 King George's Road just off the Kingsway. I claimed that it was easier to marry her than to commute across the city from Moore Park every time we went on a date.

Her father was Dr. Walter Reed Fletcher. Walter was born the 14th of October 1901 in Toronto. His father was William John Fletcher who was born in London the 28th of June 1860 'within the sound of Bow's Bells'. He immigrated to Canada with his family in 1870. He was also a physician who graduated from the University of Toronto (although his son, Walter, was a McGill grad). William John initially lived with his father who had immigrated at the same time. They were originally at 23 Esther Street in Toronto and subsequently moved to 203 Euclid Avenue. It appears that William John's father, also called W. J. Fletcher, had been a carpenter with a firm called Power & Gilchrist.

William John Fletcher married Ellen Reed on the 28th of December 1892 in the township of Erin in Wellington County. Apparently Dr. William Fletcher was one of the founders of what is now Toronto Western Hospital, part of the University Health Network. Not surprisingly this was the hospital where his son, Walter, practiced as an obstetrician. William was a Unitarian although the family background appeared to have been Methodist. William had two other children. One son, Fred, had a congenital heart problem and died early. The other was Myrtle who married to become Myrtle White.

The Fletchers: Gertie, Glenna, Grant, and Walter.

The Fletcher family had deep roots in England. William Senior's grandfather was Richard Fletcher who was married in 1828 in Horselydown, Surrey to an Elizabeth Hazard, for example.

Glenna's mother was Gertrude Beatrice Grant, always known as Gertie. She was born the 17th of July 1908 in Bowmanville, Ontario.

Her parents were Anthony Grant (1871-1926) and Annie Neilson Harper (1870-1962). They were married on the 27th of July 1897 in Toronto. Anthony Grant was killed felling an elm tree at the farm of his son-in-law William Moorey. There were some family stories that he enjoyed a drink from time-to-time although there is no firm evidence that this was a cause of the accident!

Anthony and Annie had eight children, one of whom was Glenna's mother.

The Harpers were apparently originally Irish although Annie was born in Marykirk, Scotland.

I met Glenna's grandmother, Annie, who had a wonderful Scottish accent and although a tiny woman was apparently a forceful personality. Annie was the youngest of the twelve children of William Harper and Margaret Menzies (1827-1902).

The Grant clan motto is 'stand fast'.

Small wonder that Gertie was also a determined person. She left Bowmanville to study nursing at Grace Hospital in Toronto as she described it "to meet some up and coming doctor". In this she succeeded. Walter went on to deliver over 3,000 babies and had a successful career in the west end of Toronto. They were married June 22nd, 1931.

Walter and Gertie loved their golf although she beat him regularly at the game. They had a summer cottage at Big Bay Point on Lake Simcoe called Bedside Manor. In their later years they wintered in Florida.

They became good friends of Finnie and Ruth after Glenna and I were married in 1957.

Walter died the 28th of August 1985 and Gertie passed away on the 7th of March 1987.

<u>Glenna and George:</u>

Glenna continued to teach after our marriage but added a somewhat unexpected addition to our family – a dog. We were barely back from our honeymoon in New York and Washington when she announced that she missed her old Pekinese called Kimmy. Despite the fact that we were living in an apartment at 131 Wilson Avenue we were soon dog owners.

Sunny was a small Pekinese whose full name was Suyan's Sun Yat See (a slight corruption of the name of the well known founder of modern China, Sun Yat-sen).

Glenna's interest in animals ran deep. As we will see she shortly established a kennel.

Her teaching career ended with the arrival of our first daughter, Vicki Elaine, on the 24th of March 1959. This led to the need for a house and we moved to 181 Burbank Drive in Bayview Village. The Italian builder of this new house had apparently constructed it over an old stream and we soon found that we had more running water in the house than anticipated. However this was not to be a problem for long.

My career at IBM had already taken an interesting turn. I had completed my IBM training before the company had actually brought its first computer into Canada.

An IBM 650 was brought in for demonstration purposes to the King Street office in 1957 and I seemed to get caught up in the new computer side of the business. I had taken some training on the 650 while in Poughkeepsie and this led to my being diverted to some of the major potential computer accounts in Toronto.

Specifically, in addition to being the Sales Representative at Kodak, General Electric and similar fairly large companies, I became the Sales Representative for Avro Aircraft and Orenda Engines. I was responsible for the installation of an IBM 650 at Orenda Engines which was used to design the Iroquois Engine intended for the CF105, better known as the Arrow. I was also responsible for an IBM 704 at Avro Aircraft that was instrumental in designing the Avro Arrow, the Jetliner and the super secret Aerocar.

While I was the Sales Representative, I was ably assisted by IBM Systems Engineers who were really the technical experts. My sales career however exposed me to a fascinating array of individuals including Jim Chamberlain, the Chief Engineer for the Arrow and Harry Keast, the Chief Engineer for the Iroquois Engine.

Harry Keast was an internationally renowned Aeronautical Engineer who had written a report during the Second World War that justified Britain getting into the jet aircraft business. Prior to his paper it had been assumed in Britain that jet engines while theoretically possible were too inefficient to be practical. He demonstrated that such planes had to be flown at high altitudes where they would in fact be very effective.

Needless to say my first meeting with Harry as a young salesman made me slightly nervous. I should point out that at IBM we had a strict dress code. Blue suits, white shirts and a fedora. As my first call on Harry was in the winter I also had a blue overcoat and especially for the occasion had purchased a white silk scarf with a silk fringe. I had paused in the lobby for a 'nervous pee'. On arriving in Harry's office he politely offered to take my coat. When I went to remove the scarf I found that I had firmly zippered the fringe in my trousers. I remarked to him that "they had never told me what to do in a situation like this at IBM sales school". We subsequently became very good friends.

The last time I saw Harry, he was sitting in his office holding pieces of Iroquois Engine that had been cut up with welding torches on the orders of the Government as part of that tragic end to one of Canada's most prestigious projects.

Without giving a history of computing in Canada, it is worth spending a moment describing the computers of that early era. Following several special purpose computers that IBM had developed such as the Naval Ordinance Research Computer (NORC), IBM answered the development of Eniac by producing a tube machine called the IBM 701. This was designed for scientific computing. The IBM 704 installed at Avro was the next generation. The principal change was that instead of using electrostatic (TV tube) storage, the 704 had pioneered magnetic core memory.

The 704 was one of the fastest scientific machines produced at the time working at a dazzling speed (for those days) of 40,000 instructions per second. Contrast this to the speed of an average desktop today.

The 650 described earlier had an even more primitive memory system. Instructions were stored on the surface of a magnetic drum necessitating the programmer to be very careful where on the drum he or she placed the instructions, i.e. if it took an ADD instruction a millisecond to execute, by that time the drum had rotated so that the next instruction to be used would be several spots further around the drum. Programmers today could have little conception of what it was like to program a machine as opposed to programming an operating system.

During my Toronto IBM days, I also installed one of the first new mass storage computers called an IBM 305 Ramac at White Hardware. This machine had the first disk storage. It had 50 disks, each about 18 inches across and only one access arm for the whole array. Despite its physical size it only held 50,000, one hundred character records. However it was a start in internal stored data banks as opposed to those held externally on magnetic tapes or even on punch card files.

In any case, the axe fell on Avro and Orenda when Prime Minister Diefenbaker cancelled the Arrow on February 20th, 1959. I was at Avro on that fateful Friday morning and somewhat lamely observed that I was only person on the premises who still had a job at the end of the day. Even that was a bit precarious.

IBM equipment in those days was leased and could be returned by the customer if they could no longer afford it as was the case with Avro and Orenda. I became somewhat notorious at IBM for having to absorb the largest sales 'charge back' in IBM World Trade history.

My immediate challenge however was trying to help my many friends at Avro and Orenda obtain positions. I worked with Art Downing, Chief of Computation for Avro and John Duggan at Orenda to try to line up interviews for the very talented computing engineers. However my career was about to take another change.

My experience with very large computers caused IBM to transfer me to Ottawa to install an IBM 705 Model III at what was then the Dominion Bureau of Statistics (now Statistics Canada). This was for the 1961 Census.

The 705 was the state-of-the-art computer for data processing designed to handle alpha numeric variable length records as opposed to the high speed 704 with an emphasis on fixed length floating point calculations.

Unlike the 305, the data was stored on magnetic tapes that read and wrote at 15,000 characters per second.

IBM had developed a special Document Scanner to read the census documents directly to tape. The census forms were huge and must have been very awkward for the census takers to use. However the system got the job done. It was my responsibility to ensure that this system was installed and operational in time for the Census.

The machines of this era were huge and were all tube based. Although they had back-up power, power failures were not uncommon and one had to run around the room to open the panels so that the tubes would not burn themselves out in the event of a power failure.

But back to the family. After only nine months in our new if somewhat leaky home, we had to move to Ottawa. Glenna, Vicki and I arrived in January 1960 and stayed temporarily at the Lord Elgin Hotel. Glenna was horrified to watch snow ploughs working throughout the night outside the hotel trying to remove huge drifts.

We rented a half double, as they were called in Ottawa, at 674 Hastings Avenue in the south end of Ottawa. By spring people surfaced from behind the snow banks and Glenna, Vicki, Sunny and I began a new life in the nation's capital.

On the 4th of June 1961 we had our second (and last) child, Lori Ann.

At this point Glenna and I had decided that 2 children were enough. We did, however, feel some obligation to support less fortunate children in other parts of the world. Through the Unitarian Service Committee in Ottawa, we arranged for the support of Chul Kim Ku in South Korea and Richard Mduli in Swaziland. We carried on a lively correspondence with them for many years.

We moved a lot in Ottawa, first to 27 Fairfax Avenue just off the Driveway and then to 15 Sycamore Drive in the newly developed Lynwood Village built by Bill Teron in the west end of Ottawa. We moved from Fairfax when we found that the house was infested with 'ladder beetles'. We finally settled at 3332 Riverside Drive near Revelstoke Drive on the Rideau River. This move turned out to be a good location for a subsequent career move.

Vicki and Lori received most of their primary education at Bayview Public School and then at Brookfield High.

Glenna now expanded her kennel which she had named Four Halls following the possible derivation of the Fierheller name as mentioned earlier. It was registered in 1966. Her first breed was the Bull Mastiff. These massive animals which could weigh between 125 and 150 pounds had litters of 10-12 puppies. As the 'kennel' was actually in our basement, I was somewhat apprehensive about having ½ ton of dog running around the house after a litter. I also discovered that they both ate and disposed of copious quantities. The Bull Mastiff experiment was short lived.

She next moved to the other end of the dog world in terms of size. She decided that Maltese were better designed for a house kennel. This was a very happy experience which led to Four Halls becoming one of the top Maltese breeders in Canada with 30 Canadian champions and 19 American. There were 17 Best in Show winners while Glenna had the Maltese in these earlier days. Vicki now carries on the Maltese part of the kennel with equal success but that is getting ahead of the Ottawa saga.

I was starting to spawn a parallel career path. While in Toronto, for example, I was a member of the Toronto Junior Board of Trade. One of my projects there was to Chair the Events Committee for the Miss Grey Cup Contest. This of course was a young man's dream although Glenna was somewhat skeptical. Her skepticism turned out to be right. I thought that being responsible for a dozen beauty contest entrants would be an ideal way to spend the week. I had not counted on their mothers being there as chaperones. The girls were reasonably easy to get along with but the mothers were close to impossible.

At one point Miss Ottawa woke us up in the hotel in the middle of the night to report that she could not possibly continue to participate in the contest as she had lost her lucky teddy bear. We tried to figure out what might have happened to the stuffed animal and concluded that it had likely been picked up accidentally by the maids when they changed the beds. We spent the middle of the night in the basement of the Royal York Hotel plowing through laundry. Fortunately we found it and the contest continued.

While at IBM in Ottawa I volunteered to set-up one of the first Customer Education Centres outside the main centre in Toronto. There was a real need in Ottawa for training in leading edge computer concepts and in particular the use of high level languages such as Cobol and Fortran. I rented space at 74 Albert Street, talked the Systems Engineers into organizing the courses and even designed a brochure with blue and gold type to try to give the new Centre a proper aura. It filled a great need and was a terrific sales tool.

I also established the IBM Social Club in the Ottawa office holding regular office and family affairs.

I also Chaired the Ottawa section of what at one time was the Computer Society of Canada, now known as the Canadian Information Processing Society.

However my parallel career really got started when IBM 'volunteered' me to be their representative to the Ottawa United Appeal as a Loaned Executive. There under the tutelage of one Rear Admiral Horatio Nelson Lay who was the Campaign Director I started a volunteer career with the United Way that has lasted over 45 years. I rose through the ranks to Chair the Loaned Executive group and ultimately to Chair the entire Campaign in 1971.

Back at my regular IBM career with the 1961 Census behind me, I became involved in other large government computer accounts. In those days, Ottawa was largely an IBM town. It was to become even more so.

The breakthrough came when IBM announced the System/360 on April 7, 1964. I was asked to be the Master of Ceremonies at the Ottawa launch held at the Recreational Association Auditorium of the Federal government on Riverside Drive. Again without going into tedious detail, it is important to understand the impact of this announcement. The System was designed to integrate the previously separate approach to computers that as noted earlier had led to specialized systems for scientific computing or data processing.

It was also revolutionary in that it was an upward compatible system with Models 30, 40, 50, 60, 62 and 70, each with increasing speed and capacity. The aim was that there would be one operating system called the Operating System which would work on all Models – a bit ambitious as it turned out but a remarkable concept all the same.

The announcement was to have a huge impact on the next phase of my IBM life.

I became responsible from a sales standpoint for the installation of the first System/360 in Canada – a Model 40 at the National Research Council. In fact this was the first /360 installed outside the United States. Dr. Stuart Baxter who was in charge of the program was not the easiest person to deal with but we became good friends. This pioneering installation was anything but smooth – the operating system lagged in development although the hardware operated quite well. I had a wonderful team of local Systems Engineers who wrote an interim Job Control System so that the computer, even without its final operating system, could run a series of programs without interruption.

In the meantime, IBM consolidated the top end of the /360 line by replacing the Models 60 and 62 with a new Model 65 and the Model 70 with the Model 75. One of the IBM Systems Engineers, later to become the Founder of the Computer Faculty at the University of Waterloo, Wes Graham, noted that this latter computer was so fast that we would never see one in Canada. This was one of the very few bad predictions he ever made as we shall see. He himself later installed a Model 75 at the U of W.

I was also the Sales Representative (although now promoted to Marketing Manager) for another early installation of one of the /360's at the University of Ottawa. Mike McCracken, who later went on to found Econometrica, was in charge of that installation.

I mention this because one of the applications that were being run by the Federal government using time at the University of Ottawa was a project called the Canada Land Inventory. This leading edge project funded by the Agricultural and Rural Development Agency (ARDA) was led by Dr. Roger Tomlinson and it required large amounts of computer time. The intention of the project was no less than to computerize the mapping of all the resources of Canada. A series of digitized maps covering such capabilities as the current use of agricultural lands, the best use, mineral resources, population, topography, rail lines, etc. could then be electronically overlaid to get answers to questions such as 'how much Class 1 Agricultural Land is there in a given area that is within 5 miles of a rail line'. The ARDA Project as it came to be known was being programmed under contract by IBM.

Two of the Systems Engineers working on the Project were John Russell, Senior Systems Engineer and Guy Morton, Systems Engineering Supervisor. The System was extraordinarily complex and involved pushing the frontiers of the mathematics of mapping. IBM had developed a unique scanner to digitize the manually created maps.

This Project would also have interesting career implications for me.

In the meantime, the Federal government had decided to establish a Central Data Processing Service Bureau (CDPSB) to provide leading edge development capability to all departments. However, the individual departments would continue to have their own computers for dedicated applications. The win would be critical to IBM as whichever computer manufacturer won this developmental project would have their system used for new program developments for many government departments and this would influence other sales.

I was asked to lead the team to try to win the business for IBM.

The System/360 was still in its early stages of development at this time and the Operating System was still fragile.

I assembled the best team of Systems Engineers I could find and we put a 'full court press' on the application. The Ottawa IBM team won the day and the CDPSB under Jim Radford came into being in the basement of the old Confederation building on Parliament Hill. This was in 1965 and the bid was won by a large System/360 Model 65.

We did not win all the business at that time. The newly established Canada Pension Plan was being set up by the Treasury Department and they elected to go with a more mature Burroughs System. The Federal government was ready to gamble on a leading edge computer for the CDPSB but understandably was not willing to risk a relatively unproven system on the CPP.

My involvement with the top end of the IBM line and having the Federal government as both a prospect and a client led to my being exposed to some very proprietary IBM developments. As other manufacturers tried to knock IBM off its dominant position worldwide, IBM began bidding some 'paper machines' called variously System/360 Models 90, 91 and 95. None of these machines actually made it into commercial production but they were useful to keep the competition off guard. This was the same process that National Cash Register had used decades earlier. NRC was where Thomas J. Watson Sr. first worked. In both instances such behavior led to investigations by the U.S. Federal government for anti-competitive practices.

As Ian Sharp, founder of I. P. Sharp Associates noted many years later, I had gained the dubious reputation for selling machines that never existed!

A number of these events now came together leading to yet another totally unplanned career change for me.

Systems Dimensions Limited:

The Ottawa Scene in the 1960's

The three original conspirators, John Russell, Guy Morton and I, were all part of the IBM office at 150 Laurier Avenue West in Ottawa. For some reason that office spawned a number of entrepreneurs who went on to interesting pursuits all over North America. Jay Kurtz was the first to recognize this phenomenon and actually did a study of where the 'graduates' had gone. Jay himself went on to found a lively consulting business operating out of Florida.

Just to mention a few, one could recall Olie Swanky who became President of Greyhound Computers for North America, or Ray Hession, who became President of Central Mortgage and Housing Corporation, or Jack Davies who became President of Systemhouse, or Glenn McInnes who started Alphatext, or Dave Carlisle who started Infomart.

Perhaps it was the sense of wheeler dealmanship that we picked up from Al Hewitt, the Branch Manager for the early part of the 60's. Al's philosophy was always to encourage his group to live well. He reasoned that once people got a taste for fine wines and big cars, there was no turning back – just the kind of free enterprise thinking that this young group of college types needed and which they added to the already exciting new world of computers.

As noted, I was the Data Processing Manager responsible for the ARDA project and many other Federal government accounts.

We all felt that the computerized mapping project had a great future but it was clear that IBM did not see it as a priority and were frankly concerned at the mounting costs. It was clear to us that what the project needed was a concerted effort for about a year and it dawned on us that we were the right people to do this on behalf of the government.

The whole plot was hatched in the Colonel By Lounge in the Chateau Laurier one evening in late January 1968. John, Guy and I decided we would leave IBM, form a small consulting firm and offer our services to the government to complete the ARDA project. We reasoned correctly that IBM would be delighted to get rid of the cash drain, and the government would be equally pleased to get the project quickly completed.

The next morning the idea still seemed good and we decided to give it the acid test – we discussed it with our wives. They told us we were all out of our minds, but having got that clearly on the record then helped us choose a name for this new venture.

SDL is Born

Many of the good names had already been taken. We could not call ourselves Systems Corporation or anything that general. We felt that as we were pushing into new computer fields we needed something that would emphasize the new dimensions of our project. Hence the choice of the name Systems Dimensions Limited. We had some trouble getting the name registered as there was another company called Dimensions Limited, which we gathered built door frames! However, the name stuck and SDL was underway.

The company was incorporated on March 14, 1968. Our parting with IBM was very amicable, although it was considered somewhat unusual for people to leave IBM in those days. As it turned out, IBM would be a major beneficiary in a way we had not even contemplated ourselves. We handed our resignations to Bill Moore who was President of IBM at the time and was later to become associated with SDL as Chairman of our Executive Committee.

We had decided that what we really wanted was the flexibility of being on our own and not part of a huge organization. As there was no immediate thought of expanding beyond the initial trio, we decided we may as well have whatever titles appealed. I became President and John and Guy each became Vice Presidents of a firm with no employees. This gave us a sense of power.

What gave us the money was the contract with the Federal government to provide consulting services to the ARDA project.

Guy and I planned to spend most of our time on this project while John would spend only a portion of his time, allowing some hours for the development of the new projects we expected to get into after the ARDA job was completed.

While at IBM I had sold a System/360 to the University of Ottawa. We selected this as the machine to use for testing and our operation was underway.

I should pause here to point out two important factors. First, we did not leave IBM with any idea of starting a computer services organization. Secondly, the Federal government through its promotion of a leading edge project in the form of the Canada Land Inventory had given a new company an opportunity to start.

The Summer of '68

With great confidence we rented space, bought office furniture and generally proceeded to act like a huge corporation. We chose the top floor of the newly built Place de Ville, which was Ottawa's most prestigious new office building. We had a view over the Ottawa River and the Gatineau Hills. Shortly thereafter we hired our first employee.

Hazel Martin had been my secretary at IBM and she joined us in what we later found out was going to be the start of our return to being part of a big company once again. That start came more quickly than we could have realized.

The credit for being the real innovator in the new computer services field goes to Warren Beamish and Bob Horwood who a few months earlier had started Computel in Ottawa. Computel had elected to use Univac equipment on a Remote Job Entry basis. Their belief was that there was a market beyond the interactive time sharing approach that had been developed at MIT with Project MAC and subsequently had enjoyed a significant growth in the United States.

John, Guy and I shared that belief but the next impetus came not from a computer specialist but from an Ottawa lawyer and chartered accountant – Redmond Quain, Jr. In early June Red approached Doug Bailey the then Manager of the Ottawa IBM office. Red had observed the Computel experiment but of more importance had correctly read the financial opportunity that would exist for establishing a large company to do the same thing and to take the company public. Doug suggested Red get in touch with the three recent rebels. Within days SDL was to be born all over again.

Red's idea fell on fertile ground. John, Guy and I had always felt that Computel had slightly missed the market. Ottawa was an IBM town in those days. It seemed to us that the conversion of accounts from in-house computers to a large IBM Remote Job Entry facility was a natural. We knew the equipment and knew exactly how to go about launching such a program.

The feasibility of RJE was made possible by the increasing size and capability of main frames, the sophistication of multi programming operating systems, and the increasing reliability and availability of telecommunications facilities. When Red advised us that he felt the financial community could support a multi million dollar underwriting, our imaginations went wild.

While still working full time on the ARDA project during the day, we spent our nights and weekends over the next several weeks designing the ultimate in a computer service centre which we called the Systemcentre.

Our concept was to design and build a completely new type of building oriented toward this new industry.

We would buy the largest and fastest commercially available computer which it turned out was the newly announced IBM System/360 Model 85.

We would design and implement an entirely new approach to accounting for the use of time by multiple users on the same machine. Prior to this, most time sharing operations used essentially a wall clock approach, which was a very unsophisticated allocation of costs to the user. Our approach would be to monitor minutely the actual time a user tied up various resources. We would go further and so accurately analyze the way computer programs operated so that we could assist users in optimizing the way their programs actually ran.

Finally, we would pull together the best people we could find to make all this happen. Whether we liked it or not we were launched on a course that would make us once again part of a large organization of the type we had just left.

All this was put together within about four months of the time we had left IBM.

We drafted a mini prospectus and added up the costs which we now estimated to be a staggering $14 million.

Our wives simply shook their heads and decided that they had accurately assessed the situation earlier when they concluded we were all mad!

What's a Million?

It is always better to be lucky than brilliant. Red had undertaken to contact an underwriter and remembered a classmate of his, Ross LeMesurier, who he found out was now a Senior Vice President with Wood Gundy. Ross is one of those who combine energy and imagination in boundless quantities. After reading our background paper he flew to Ottawa to meet these four people with their multi-million dollar idea.

It was the start of a period of the most intense activity any of us could remember. It was strange in a way that Canada's largest underwriting house would be willing to gamble on a small group of people with no track record in business and essentially no money. Perhaps it was because Ames & Company had taken Computel public and Wood Gundy felt the need for rounding out their portfolio for clients with something that was Canadian and in the high technology business. Whatever the reason, the association was a lucky one and Ross was in invaluable ally in moving us from an idea into reality.

We had also approached Bankers Trust from New York as we were frankly not sure that Canadian financial institutions would recognize what we perceived to be the great future of the new technology. We soon settled on the Wood Gundy approach however, and this Canadian association itself likely helped our initial sales with the Federal government.

We set ourselves an unbelievable schedule. If the financing could be arranged, we committed ourselves to have the entire operation running by June 30th, 1969 – at this point less than nine months away.

This meant we had to start immediately hiring staff, committing an order to IBM for what would be the first commercial installation of a Model 85 and of course we had to buy land and start the building. For a second time our faith in the Canadian financial institutions was to prove well founded.

The money from the underwriting would not likely come until February 1969. We needed bridge financing for what had now escalated to a $17.5 million deal. Our bankers were not the Toronto-Dominion but Dick Thompson, the newly appointed General Manager, heard about the potential underwriting through Wood Gundy and flew to Ottawa to meet with me.

Dick's opening statement was, "We would like to be your bankers. What do we have to do to get the business?" I took a deep breath and said, "How would you like to lend us $1 million?" Dick's reply was, "Fine, What else?"

The Toronto-Dominion ended up advancing us about $1.25 million and in September we started hiring, ordering equipment, and designing the Systemcentre.

Perhaps it was the confidence we exuded from our offices atop the Place de Ville. More likely it was just the happy combination of circumstances of an up stock market, an exciting idea, and some imaginative Canadian financial organizations that made this possible. The enormity of the gamble we were taking seems more shocking in retrospect. At the time we were so caught up with the fascination of the project, we never stopped to think about it.

The People

The project attracted people. Many came from IBM where their backgrounds would help our tight schedule. We quickly added Frank Van Humbeck who had left IBM to move to Domtar in Montreal. He was to prove an invaluable participant in the development of the new accounting system which we had dubbed Accountpak. Don Pounder, an Advisory Systems Engineer with the IBM Ottawa office joined and immediately took on the task of overseeing the building construction – not something his background would have led him into under normal circumstances. But then these were not normal circumstances. Brian Greenleaf, Charles Ploeg and others from the Ottawa office strengthened the technical team. We recruited Jack Kyle an IBM Toronto Branch Manager as the General Manager. He quickly recommended Bill Beairsto as the Director of Marketing.

We recruited widely from the University of Waterloo where Paul Cress, one of the developers of WATFOR, was to add immeasurably to the team.

Red recommended Jim Keogh as our Controller. Jim was of Irish extraction and had ten children when he joined us with more on the way – an odd qualification for a Controller but we hired him anyway!

By the end of the year we held a party and discovered we had over thirty employees. Hazel took one look at the roomful of people and observed, "I thought it was going to be just a small group." She said, "Now we're getting as big as the IBM office itself."

In the meantime, with the help of Ross LeMesurier, we were assembling a unique and innovative Board of Directors. Walter McCarthy, Senior Vice President, Finance of Sun Life, accepted our offer to join the Board. Walter was to play a major and unforeseeable role in the company some years later.

Jim Tory, of Tory, Tory Deslaurier and Binnington, demonstrated that lawyers can think like businessmen, and was to help greatly in the times ahead.

Philippe de Gaspé Beaubien, President of Telemedia of Montreal, was equally fascinated with the new venture. Philippe had been the General Manager of Expo 67 and quickly suggested that the new company would need an entirely new and innovative communications program.

He recommended the firm of Brake, MacDonald, Paine & Watt, who had done imaginative work for him. This group set about designing the logos, choosing colours, and creating a quality image that the fledging company so clearly needed.

Professor Wesley Graham of the University of Waterloo had been a long time friend at IBM and was himself one of the computer pioneers in Canada. His addition to the Board provided an independent technical input.

It was a people business. We were pulling together the right people to do the job.

The Systemcentre

We had acquired 3½ acres of land on Brookfield Road adjacent to the new airport parkway. It was essentially a residential area but we placated our neighbours by hiring Murray and Murray, the noted Ottawa architects, to design for us the unique and attractive building we felt the Systemcentre should be.

The difficulty was the size of the computer configuration we had selected. The Model 85 was the first water cooled machine in the IBM line. Its massive memory, for those days, and its many high speed components meant that we would need a building ranging over most of the property to hold all the devices. At one point John said, "Wouldn't it be nice if we could simply hang some of the

Fly-eye view of the Systemcentre lobby and computer floors.

components from the ceiling?" The idea turned out to be just what we needed. We took the components which required very little access and designed a pedestal that rose above the main computer floor on which these components could be placed. The cables ran down the support columns and out to the printers, discs and tape drives and other units requiring operator attention on the main floor.

The design was striking, leading to comments that it looked like a computer altar! This was appropriate, as by this time we were praying that everything would be on schedule for the June opening.

We started construction in January 1969 in the midst of an Ottawa winter. Incredibly enough building was completed so that the computer equipment could be moved in by the end of April. We could not get delivery of the Model 85 in time and installed an interim Model 65 to complete the testing of the complex Accountpak routines and other innovations.

One of our great finds had been Norm Williams who joined us as Director of Administration the previous Fall. Norm managed to think of the myriad of details that had to be looked after in parallel with the building and computer installation.

Supermoney

By February of 1969 the underwriting had been completed and the company went public. The stock was substantially over subscribed. The offering was in the form of units with a $1,000 debenture and seventy-five shares at $10 each. The shares rose the first day to over $20 and by the second day were trading over $25.

The founders each had over 120,000 shares for which we each had paid $535 in total. The Alberta Securities Exchange Commission referred to these founders shares as being "unconscionable". We were all instant millionaires.

This was true Supermoney in the Adam Smith sense (the author of a book by that name). Several of the habitués of the Skyline Men's Club which we also frequented suggested kidnapping us. They of course had not read the fine print in the underwriting agreement. We were in escrow for five years. That is the downside to Supermoney.

Success and Survival

The Systemcentre opened on schedule on June 29th, 1969. We thought it would add to the facility to have an extremely attractive Receptionist in our beautiful new lobby. The young lady we hired looked something like Marilyn Monroe but the idea backfired as most of the young men crowded around her desk rather than using computer time. As President, it was up to me to fire her. She came into my office with a low cut blouse and a mini skirt and when I delivered the bad news she asked "but Mr. Fierheller, what is wrong with me?" I could not think of very much but tried to explain that with her obvious talents she might have a better career elsewhere. This was one of the challenging tasks a President must handle!

One Ottawa writer gave a glowing account of the facility under the heading, "The Computer Palace, where the Client is King." There was one thing we had forgotten however – there were no clients.

We had expected to swing the sizeable ARDA computing contracts from their temporary home at the University of Ottawa to SDL. The government had let a tender for this but we had no sooner opened the doors of the Systemcentre than the government announced one of its periodic major cutbacks. The ARDA contract award was postponed. We had made the mistake of becoming too dependent on the Federal government marketplace for our initial launch. Suddenly it was simply not there.

As part of the underwriting we had obtained an outside consultant's report from Booz Allen & Hamilton. They had recommended an adequate cushion of money to carry us through the build up of clients. Neither they nor we had anticipated this sudden change in the Federal government scene.

We had to act quickly and we did.

The AGT Agency Arrangement

AGT Data Systems Limited was a company started in Toronto by Gerry Wanless and Ted White. These were two fine marketers with offices in the two largest non-governmental areas in the country – Toronto and Montreal. SDL made arrangements with this company to market the new services in these cities to take up the slack left by the faltering Federal government market.

As President of SDL, I always endeavored to assist in the closing of new business. One of our prospects was Hiram Walker. After our sales group had done their job, it was left to me to close the sale at a luncheon at the Green Valley Restaurant in Ottawa. Without thinking, at the lunch I ordered my usual very dry Beefeater Martini on the rocks with a twist. The sales group looked on in horror as Beefeater was clearly not a Hiram Walker brand.

It took them several months of additional work to finally close the business.

Gary Hughes headed the SDL operation in Toronto and Dave Carlisle did the same in Montreal. Gary was later to join SDL as Vice President, Marketing. Dave Carlisle later moved to Datacrown where our paths crossed once again. But this is getting ahead of the story.

The Toronto and Montreal offices started to produce commercial clients but other problems were on the horizon.

I Hope You Have Good Relations With Your Bank

It was becoming obvious that with the slower sales build up, the costs of accessing the Toronto and Montreal markets remotely, and the need for more advertising and other expenditures, we were running out of money. Price Waterhouse, our accounting firm, came to see me one day in early 1970 and said, "I hope you have good relations with your banker." Our cash flows indicated the need for another $1 – 5 million before we could comfortably arrive at a breakeven point.

Once again Wood Gundy rose to the occasion. By this time the bloom had gone off the stock market. In fact, had we attempted such a huge underwriting even six months later than we did, the program would never have got off the ground.

We therefore decided to try the private placement route. We felt that if we could get 12-15 organizations to contribute perhaps $100,000 each the problem would be solved. This was easier said than done.

Although a number of Canadian companies were willing to take up the shares, bonused by some founders' shares, we had to go offshore for some of the money. Ross and I did a whirlwind tour of Canada and ended up in Zurich where the Union Bank of Switzerland added the necessary final financing.

As it turned out, we had panicked a bit too early. By the end of the first year sales were up to $1.3 million and the company could have scraped by without the additional financing. However, by now we had new uses for the money as we were becoming acquisitive.

SDL Informatique

We had decided that if we were going to properly operate in the province of Quebec, we needed a Quebec base of business. This as much as anything else led us to break off the agency agreement with AGT. We took over the operations of the Toronto and Montreal offices. Meanwhile, I had been endeavoring to sell the Quebec government on using computer services. They replied that unless we had a computer presence in Quebec City there was little hope.

I approached Pierre Cote, then President of Laterie Laval. Pierre had a small company in the software development business called Infomatel. He agreed to sell this company to SDL and joined the SDL board. We then established a Quebec corporation called SDL Informatique to run the Quebec and Montreal offices.

The American Dream

By the end of our second year sales had increased to $4.1 million and with this our confidence was rising. We looked south to the huge American market.

As we knew next to nothing of the area we hired a consulting firm to help us find an appropriate partner. This organization had an associate called Sal Salerno who took charge of the search. I can remember one occasion when Sal took me to one of his favorite Italian restaurants in New York. I had the vague feeling that at any time the Mafia were going to come in with submachine guns! However, undaunted, we pushed on in the American market. Sal in turn retained the services of Julius Honig. Julius and I had known each other when he had been associated with IBM's abortive System/360 Model 67 project and I had been trying to sell one of these to the National Research Council. Big Julius, as we called him did indeed find an organization based in White Plains, New York who agreed to open offices for us in Boston and New York City.

The deal was concluded with EDP Industries at the Broad Street Club in New York with all the solemnity that a Wall Street setting should have. Little did then President Brian Satterlee and our new associates know that the prior evening the founders had celebrated this new event by riding down the streets of New York in the stage coach that the Cattlemen Restaurant used to run!

We were now an international company and subsequently opened up in Washington in the hopes that we could transfer some of our government application expertise to the U.S. capitol. We subsequently bought over the EDP Industries operation and created SDL International.

An Information Company

To this point SDL had been primarily a computer services organization. Then Norm Williams made the mistake of reading Drucker's "The Age of Discontinuity" which predicted that we were entering a new age where the information industry would be king. All of this took place at around the time when the Federal government was running the telecommunications enquiries and publishing books like "Instant World".

We felt we had a role to play as a multi-faceted information provider as well as just a sophisticated computer services company.

We were about to learn the price of diversification away from things we knew well.

But we were also caught in a trap of our own making. After the initial explosion in the stock price, the expectations for the earnings of the company had escalated to match. We were now scrambling to find things we could add to the company to keep the earnings growing at the rate the financial community had come to expect.

One of our many ventures while at SDL was to investigate the use of large scale computers for language translation. We came across an organization based in La Jolla, California. At the instigation of the U.S. military, they had developed a very advanced program to translate English into Vietnamese. The intention was to use this to translate maintenance manuals and other documents as part of the Vietnamization of the armed forces toward the end of that unfortunate conflict. The idea was great but the group was off the wall. At one point during a meeting they politely excused themselves, went over to the corner and stood on their heads. They explained that yoga helped to clear their minds for this difficult project. It also cleared us out of the organization.

Softwarehouse Limited

In line with our planned development into an information company, we acquired a 23 person software development organization headed by Jack Davies. Our feeling was that if we could help clients develop new applications we would in turn sell more computer services. In fact, this proved to be accurate. The Softwarehouse division grew to over 100 people and soon was contributing considerable revenue. By mid 1972 SDL's revenue was at $5.8 million.

However it was time to move on.

Systems Research Group

A Toronto based organization called SRG was a user of the SDL system. They were a very profitable organization specializing in analyzing information needs and developing systems for the education and health care industries in particular. The firm was headed by Dr. Jack Levine and Professor Richard Judy.

The main product offered by this company was called CAMPUS which was a planning, reporting and analytical tool for educational institutions. It had been developed with government funding and the ownership of the product was not entirely clear. However, what was clear was that SRG were the only people interested in trying to capitalize on this in a commercial sense.

Jack Levine was tall, good looking, personable, brilliant and wealthy. Naturally his peers at SDL viewed this new arrival with some reservation!

We paid for the company largely with SDL stock on an earn-out basis. The association started well enough although at one point Paul Cress was heard to remark, "I hope

George remembers that we are still back here at the Systemcentre pedaling the bike". It was in fact true that the computer services business itself was growing at a good clip and some of our new adventures were starting to look like diversions.

With the help of Jack and his group we created SDL 100. This was the long range plan for SDL as an information company and projected our growing to $100,000,000 by the early 80's. As it later turned out SDL, when merged with Datacrown, came close to achieving that figure on schedule but we were a long way from there in 1973 with revenues of about $12 million.

Jack Levine even got an Ontario licence plate with SDL 100 on it.

I used my rapidly developing ability to create a logical looking organization out of utter chaos and divided the operation into three groups:

SDL Computer Services Group comprising such things as computer services, facilities management, software products and application management.

SDL Systems Research Group which would work on information systems, planning and budgeting systems, simulation models and related advanced applications.

SDL Institute which under the direction of Professor Judy would look at policy research, cost benefit analysis, economic research and the new field of social studies.

We began to think of SDL as being the information think-tank for Canada.

Within the Systems Research Group we established three major thrusts in education systems, health systems and general systems.

I had finally convinced myself that we now had the correct format for the organization as we could analyze the broad social or policy needs of organizations through the Institute, develop the information systems through SRG, write the computer software through Softwarehouse and run the whole thing on the SDL remote job entry computer service.

There were just a few things I had forgotten. The company had now grown to over five hundred people where the average age was under twenty-nine and the educational level astoundingly high – over half were college graduates. Co-ordination was becoming a problem. We were moving too fast to have learned to work together as a group. Hazel Martin's fears of 1968 were becoming a reality.

Shortliffe & Associates

In May 1973 the health care division of SRG acquired Shortliffe & Associates. This was really the services of Dr. Ernie Shortliffe who had had extensive health care institution management experience with Extendicare and elsewhere. One of Ernie's projects was to come up with a hospital version of CAMPUS. This was irreverently referred to as "CAMPUS in Green" but its more formal name was MEDIC. However, as computer systems were not really Ernie's field I discovered that much of his time was being taken up with contracts such as hospital facilities management – not quite the information company approach I had envisaged. The organization was getting a bit far afield.

Anathon Computer and Educational Systems Inc.

The next acquisition was SRG's venture through the educational division into student reporting. This was a field in which we had some experience but we concluded that the best way to promote this was to acquire a New York based company with reputedly the best school accounting and student reporting system around. By this time I was getting quite used to this acquisition game and acquired the company on the ultimate earn-out, i.e. nothing down with the owners being paid only on performance basis over a number of years.

I recall attending a University of Toronto seminar on acquisitions and mergers and being asked at the opening session if I minded their using Anathon as an example of the perfect acquisition. They even asked if I would return the next year and lecture the group on how it was done.

I was never to have the pleasure! What I had not realized is that however clever the earn-out may be you still own the company. It quickly developed that the system was not everything it was supposed to be. The first installation was only a success because the developers hand held the whole operation. Unknowingly we used this model to sell the city of New York on a student reporting operation.

It was typical of our bravado to take on all of New York at once. It took months of our best people's time to get the whole thing into operational shape.

The problem grew worse when the city of New York was finally acknowledged to be bankrupt.

With about a million dollars in receivables and a limping system we astutely got out of the whole operation by selling it back to the original developers for a dollar. Fortunately we eventually collected the receivables but the University of Toronto looked elsewhere for a lecturer on acquisitions!

Ottawa Cablevision Limited

As part of the SDL 100 program we had planned to enter the Wired World. Our belief was that the new services being developed by SDL could be delivered to the home and office over cable. This idea was not new but we were going to be the first to put it into practice in North America.

To provide the appropriate electronic laboratory we set out to acquire Ottawa Cablevision Limited, the largest of the two cable companies in the nation's capital. A deal was struck with Gordon Henderson representing some of the Ottawa investors and Selkirk Communications Ltd. The acquisition was of course contingent upon CRTC approval.

We felt the marriage was a natural one that would allow us to develop such applications as teleshopping, in-the-home education and a variety of other such applications. The CRTC hearing took place before the then Chairman Pierre Juneau in June 1974. The decision has become a Canadian classic. It simply stated that such a merger for these purposes was "premature". We had to unravel the OCL deal but at about this time some other things started to unravel.

CTV Lawsuit

SDL through the Softwarehouse Division had handled the processing for a number of elections live on TV. For the Federal election in July of 1974 the CTV network encouraged us to try a very leading edge system involving extensive on-line input and multiple terminals.

"Excuse me, I think my finger is stuck!"

On the evening of July 8, I was watching the results and quickly realized that all was not well. Frank VanHumbeck called about 9:30 that evening and indicated that the input data was far exceeding the rates we had been led to expect and the system was floundering.

The next morning I was called by a reporter who advised me that CTV was suing SDL for $2.5 million. The writ was primarily a pubic relations gesture and subsequent analysis indicated that the fault was as much CTV's as ours. However, it was a jolt to our established image for technical infallibility and the jittery stock market reacted with what was the start of a long downward trend.

The suit was settled amicably out of court with an agreement that merely saw SDL not charging CTV for the services provided.

Short Sharp Shocks

Jack Davies now felt that SDL had simply grown too large and too diversified and decided to leave. I was very sorry to lose Jack, and even more so when he took about twenty-five of the Softwarehouse group with him including John Kelly, later the President of NABU. Jack went on to establish Systemhouse Limited, one of the country's most successful software development operations.

The next shock came when Jack Levine and Dick Judy who were now at the end of their earn-out period bailed out of most of their SDL stock. In the weakening stock market this put further pressure on the market for our securities. Supermoney did not look as good as it once had.

Jack Levine and Dick Judy then resigned. It was decision time for me regarding the Systems Research Group! The SRG program which had really been theirs was folded shortly thereafter. SDL was coming full circle back to being a computer services company although still strong with a development program in packaged application services.

The executive group was already changing. Red Quain had left the business in 1971 as he felt his future interests were not in management. John and Guy were no longer active in the business in a management sense although they continued on the Board of Directors until the merger of the company with Datacrown.

SDL in 1975

Despite the problems the company had grown to over $21 million in sales with a profit before extraordinary items of over $1 million.

SDL itself had often in the past been a target company for acquisition. In January 1972 Bob Scrivenor, then President of Bell Canada, came to see me with the suggestion that Bell buy SDL and use this as its entry into the computer services industry. This was a very controversial issue at the time but had some attractions. Not the least of these attractions was my perception of what it might be like working for Bell. It was a bitterly cold January day when Bob arrived in my Ottawa office. I thoughtfully suggested that if he had driven from Montreal he might put his car in our basement garage where it would be warmer. He said, "Oh, that's alright my driver will keep the heater going". Being part of a really big business had some personal appeal!

That particular acquisition did not take place but by 1977 SDL was the only large independent computer services operation in Canada. Datacrown was of course a subsidiary of Crown Life. Canada Systems Group was jointly owned by Eaton's, Stelco and Gulf. Computel had been bought by Royal Trust. Multiple Access had been acquired by the Bronfman interests.

To compound our vulnerability we had developed a very strong financial position with close to $5 million in cash. It seemed to me that if the industry was going to develop in Canada something had to be done to rationalize the then fierce competition amongst these large companies. During 1977 I had a number of discussions with others in the industry as to how this might be brought about. One of these was with Dick Taylor, President of Datacrown.

SDL and Datacrown

Although a number of industry combinations might have been possible SDL and Datacrown had very complementary markets and relatively similar approaches.

On a geographic basis SDL had operations in Quebec and the United States but was relatively weak in the Toronto area. Datacrown had strong operations in Toronto and along with SDL small operations in the West.

From an operational standpoint the computer equipment was similar and the computer accounting approaches somewhat alike. However, Datacrown concentrated on machine replacement where SDL was promoting the application development business.

The executives of Crown Life agreed to look at the possibility of a merger. But all was not to go according to plan.

SDL and Sun Life

Walter McCarthy recognizing that SDL would likely be acquired by someone brought the matter to the attention of Tom Galt, President of Sun Life. Walter declared his conflict of interest and absented himself from any discussions that might involve other suitors. By then Sun was clearly interested in the possibility of purchasing SDL.

Battle of the Giants

Crown Life made the first move with an offer of $4.00 a share. This was a reasonable premium over the depressed SDL stock price. Sun Life countered with a $4.50 offer. The Crown Life Board decided that such a bidding war was not going to be advantageous and declined to increase their offer. Sun Life set about accumulating the shares to gain control in what should have been a routine takeover. It was not to be.

The Maritime Move

In mid 1977 while this offer was being considered by the public the stock started to trade erratically. Purchases were made above the Sun Life offer. Large blocks were being exchanged of which neither Sun Life nor Crown Life claimed any knowledge. It was clear that a third bidder had entered the scene.

Shortly thereafter an announcement was made by Coastal Enterprises Ltd. of Hantsport, Nova Scotia that they had accumulated a block of shares and were about to make a bid of $4.60 a share. Coastal Enterprises was a small holding company of the Jodrey family interests. John Jodrey, who was on the Board of Crown Life, believed that Crown was letting a great opportunity slip away and decided to bid for the company for himself.

To everyone's astonishment, Sun Life who had just got the upper hand in the bidding war with Crown Life decided not to get into another such battle with comparatively unknown forces. Coastal Enterprises won the day.

ADP Systems Limited

SDL had been on a plateau for about a year during this acquisition activity. Under the new ownership however SDL launched once again on the acquisition trail. SDL had opened offices in Calgary and Vancouver but had never been active in Winnipeg. The opportunity arose to acquire a company that had been started by Dave Cortens called ADP Systems Limited. Its major asset was a contract to provide accounting services to IATA. The application was the reconciliation of flight coupons for all the airlines operating in Canada.

This acquisition was completed in 1978 and became a profitable operating western based subsidiary.

A Coastal Company

For about a year SDL operated as a wholly owned subsidiary of Coastal Enterprises. It was a somewhat disquieting experience for the employees who had been on the receiving end of an acquisition skirmish for the first time. However Coastal under the leadership of John's nephew Dave Hennigar and Bob Granger of Aird & Berlis proved to be a most understanding group with which to work.

My first meeting with them took place in the Laurentian Club in Ottawa shortly after their bid succeeded. David asked for a private dining room and we all thought it fitting when we were informed that the club had reserved the <u>Sun</u>room for us!

Datacrown Revisited

In the Fall of 1978 Coastal Enterprises approached Crown Life with an offer to sell the company to Crown with the intention of merging SDL and Datacrown. The offer was at the price paid by Coastal Enterprises.

The logic of this move which had been apparent for so long led to its acceptance by the Crown Life Board. During the Fall of 1978 SDL merged with Datacrown to form what came to be one of North America's largest computer services companies with sales in 1981 in excess of $80 million.

SDL in Retrospect

After the merger the name SDL was dropped, the staff merged with Datacrown and the computer systems were integrated. I stayed on as Vice Chairman of the Board of Directors to assist with the integration but left the management of the company in March 1979.

SDL enjoyed a colorful history. For much of the decade it had been the largest independent computer services company in Canada. The people in the company had pioneered many new concepts. Graduates from SDL have gone on to start many new and innovative corporations.

The stock made a lot of money for a lot of people and with its ups and downs cost a few people some money as well!

In the end however SDL's major contribution was not just technical or financial. Its main accomplishment was demonstrating that young people with new ideas can overcome all kinds of problems to create new industries. And we showed it can be done in Canada – a happy ending for a lively decade.

The spirit of the people involved in SDL was such that they stayed in touch for many years afterwards. In fact almost a decade after the company ceased to exist as a stand-alone organization, there was a gathering for SDL's 20th Anniversary in November 1988 that was attended by over 200 people assembled from all over North America.

The Ottawa Years

My SDL days were filled with adventure. I will only recount one not untypical incident. As noted earlier, I always seem to get into trouble in Paris. In 1970 when Ross LeMesurier and I were doing our Road Show to raise additional money, our last stop was in Paris. When we had finished our final presentation, Ross and I decided to celebrate. We were staying at the Plaza Athenée.

We tried to get into the Moulin Rouge but it was full. A young Gavroche assured us that another show around the corner was equally good. We should have known better. Before the evening was over, Ross and I had bought at least a dozen bottles of Champagne, which we shared with the show girls, the waiters and likely all their friends. We had naturally consumed our own share of this and I vaguely remember signing traveler's cheques to pay for it all with the kindly proprietor egging me on by saying 'un autre monsieur'.

We almost crawled into the Plaza Athenée at about 3:00 a.m. to be greeted by the doorman complete with his Chain of Office. We had agreed that we would meet the next morning because Ross's wife, Elizabeth, insisted that he should see the main tourist attractions in the city before we came back and she had given him a list.

Ross had lost a leg in the Second World War and had an artificial limb. He was in the room next to me and the following morning I heard a strange thumping noise. Apparently he had lost his limb a second time. We finally located it behind the sofa and once we had it strapped on we left for the tour. We hired a taxi who took us first to Notre Dame. Ross asked the cab driver to point out which side the Cathedral was on, quickly ticked it off on his list and we drove on.

The next stop was the Hôtel des Invalides. The cab driver proudly announced that this was where all the great French military heroes were buried. Ross observed out loud that "it must be nearly empty". The cab driver gave us a black look and proceeded to the Avenue de la Grande Armée. Ross commented that it "must be a one way street out of town". Ross's recollections of the French during the Second World War were obviously not positive.

At this point the cab driver stopped, ordered us out of the cab refusing even to accept payment considering the insults he had already received.

The culmination of the trip was to return to North America via New York on one of the very first Pan Am 747 flights. We flew first class in the days when the upper deck was a bar lounge complete with caviar and more champagne. There really was some fun in <u>fun</u>draising!

Even during the rather hectic SDL days, my parallel volunteer career continued. I was one of the few entrepreneurs in Ottawa in those days and was therefore invited to serve as the token business person for many voluntary activities.

The National Arts Centre had just been started and I was asked to sit on its Board and Executive Committee. I had had some volunteer experience as Business Manager for Opera Atelier, a semi-professional group that had pioneered Opera in Ottawa.

The invitation to join the NAC Board arrived by a phone call from the then President, G. Hamilton Southam. The message said that he would like me to call him. I replied with a message that G. Alfred Fierheller would be pleased to talk to him the next day. This led to an exciting few years as the NAC put on special performances for visiting dignitaries. It was also the golden age of the NAC which amongst other activities produced three full Operas each year.

In 1970-71 I was the National President of what was formerly known as the Computer Society of Canada, now the Canadian Information Processing Society (CIPS). When I decided to run for the position my friend, Wes Graham, said I did not stand a chance. CIPS was the Professional Association for elite computer people and prior Presidents had been those such as Kelly Gotlieb and Pat Hume who pioneered computing in Canada with the FERUT (Feranti U of T) Computer, Harvey Gelman and Wes himself. However, I managed to convince the voters that I was a Science graduate although I de-emphasized the fact that it was Political Science and was swept into power.

I used the time to set up the first Canadian Computer Show jointly with Maclean Hunter, started the CIPS monthly magazine and inaugurated many of the discussions on the certification of computer professionals. As the national guru of Canadian computing for the year, I was Canada's representative to the 25th Anniversary of the first electronic computer, the Eniac. The meeting was held in Chicago and I had what for me was the awe inspiring experience of actually meeting Eckardt and Mauchley, the designers and builders of the first truly general purpose electronic computer.

I had always had a strong interest in post secondary education and Carleton University was very near the Systemcentre. Because of my experience of Chairing the United Way Campaign, I was asked to Chair the Carleton University Capital Campaign. I moved from there to being Chair of the Associates of Carleton (the founding group of the University) and then became Chair of their Board of Governors from 1977-79.

Neither IBM nor SDL were ever unionized. My predecessor as Chair of the Board at Carleton had succumbed to a Faculty Union. It was left to me to negotiate the first contract. For a business person this was an exercise in frustration. Those on the other side were clearly very bright, tended to be academic nit pickers and further seemed to have all the time in the world. We finally got an agreement after I threatened to impose time clocks as I claimed they were acting like factory workers.

Working with students is always a pleasure although it too had its challenges. At one point they proposed a march on Queens Park demanding lower academic fees. I was asked to address a rally of the students and I am sure they fully expected me to tell them not to go. Instead I congratulated them on the initiative, advised that they were exercising their rights and obligations to make their voices heard and only asked that they behave as the responsible future leaders of the country by not rabble rousing but rather putting forth well thought through arguments. The then Secretary of the Board, Don McEwen said afterwards that I had "done alright".

I also served on the Board of Trustees of the Royal Ottawa Hospital which was actually a Psychiatric Hospital – probably appropriate given my apparent inability to say 'no' to any interesting idea or worthwhile charitable function.

During our stay in Ottawa, Glenna, Vicki and Lori decided to take riding lessons. They talked me into going along. I have a natural respect or perhaps disrespect for these large animals. They seem to lack an adequate steering mechanism and certainly lacked brakes.

We had an autocratic German instructor who immediately picked me out as being a hopeless pupil. The others were given horses with names like Black Knight or Charger. For me they provided a horse called Daisy.

Even Daisy proved beyond my control. Whenever it felt like stopping to eat or drink it simply went ahead. The instructor was disgusted. Finally Daisy decided she was going to roll in the dirt. The instructor suggested it might be a good time to get off the horse which I did by grabbing a fence post and pulling myself to safety.

The rest of the family rode on absolutely mortified.

Lori loved to ride and at one point took a summer job providing riding instruction at a camp south of Ottawa. Glenna was off at a dog show and I agreed to drive Lori to the camp. That morning I had my eyes examined and the Optometrist had put drops in my eyes. It became obvious that I could not see to drive and so with my usual fast reaction to any situation, called a limousine service to take the two of us to her camp. What arrived was a stretch limo with a uniformed driver in full livery. When we arrived at the camp, Lori could hardly get rid of me fast enough before her students saw her arrive in this up-market fashion.

But our Ottawa years of 1960 to 1979 were almost over. I had advised Crown Life that I was planning to leave Datacrown on March 31st, 1979. I now received a gentle prod from Glenna who suggested that it might be a good idea to start looking for a job.

With my usual casual approach to career planning I looked in the Want Ads. By luck there was an ad for the Presidency of Premier Cablesystems, at the time the largest cable television company in Canada as well as being one of the oldest. It was based in Vancouver which seemed to me to be an interesting place to live.

I took the interview and although I had no hands-on experience in the broadcasting field, I was able to use the buzz words I had picked up from my exposure to Ottawa Cablevision. I managed to convince them that I could lead a slightly tired company into the new world of in-the-home services over cable.

My announcement of our imminent move to Vancouver brought a somewhat skeptical response from the family. Glenna, Vicki and Lori had all been in Ottawa for 18 years and had many friends and connections.

Their enthusiasm was dampened even more when I reported that I had found a home in Vancouver and it was located on an Indian Reservation. I sent a picture of the Musqueum Band Office that was located at the end of our street. The sign was partly falling down. This was not a positive sales move! However the newly built house at 4184 Musqueum Drive was a hit once the family checked it out. It was indeed on Indian land which we could only lease. However the lease was for 99 years and I reasoned that this was quite long enough for me.

The house was just off South West Marine Drive, behind Shaughnessy Golf Club (which I promptly joined) and was right next to the University of British Columbia Endowment Lands. As it turned out it was also reasonably close to the airport which would turn out to be a blessing.

The Vancouver Years:

I started with Premier on April 1st, 1979 which to Glenna's relief meant I had not missed even a day's pay.

Vicki with her endearing love of animals was already enrolled in the three year Veterinary Assistant course at St. Lawrence College in Kingston. She wisely decided to complete her course but came to Vancouver for the summers.

Lori had completed her Junior Matriculation (grade 12) at Brookfield High. Although it was unusual to enroll a student at the University of British Columbia without Senior Matriculation, she was accepted in first year of Geological Science.

Glenna's Maltese had moved with us to Vancouver and she was soon active in the dog world on the West Coast.

Premier in the meantime was providing an interesting challenge. It was a public company but with strong controlling shareholders. Sid Welsh, the Chair, had been one of the Cable TV pioneers in Canada together with Ed Jarmain in the east. Sid's partner was Garth Pither. The Premier Board had such fascinating members as Frank Griffiths who owned the Vancouver Canucks, and Austin Taylor, Chair of McLeod, Young, Weir.

The company had major cable television systems in Vancouver, Victoria, Calgary, several systems in the Toronto area and even two in Ireland. However they had had a somewhat rough relationship with the Canadian Radio Television and Telecommunications Commission (CRTC) and some aspects of the volatile Vancouver community. B.C. was always known for its activists and the Premier CRTC hearings attracted many negative interveners.

Their stock was also languishing around the $9 - $10 mark and it was almost immediately obvious that the controlling shareholders really wanted to sell.

Ted Rogers had recently completed an unfriendly takeover of Canadian Cablesystems Limited and was busy consolidating this with his own Rogers Cable TV licenses.

Ted had already expressed an interest in Premier. The story of his arriving at a Premier Board Meeting with a Hawaiian Band that he had picked up on the flight out was by that time legendary. However at that time the somewhat reserved Board was not amused.

The time now appeared right for Ted to make a move and I prepared the company for possible sale. The two seriously interested parties were Izzy Asper and of course Ted. Ted's bid of $25 per share capped the Asper bid, and Premier became part of the rapidly growing Rogers Communications empire.

The former owners were delighted to get over 2.5 times the share price at the time we started discussions. The CRTC, after extensive hearings approved the merger in 1980. Ted and I were about to begin a long business association.

I stayed on as President of Premier but also became Chair of the integrated companies under the name of Canadian Cablesystems Inc. In my usual volunteer role, I became Chair of the Strategic Planning Committee of the Canadian Cable Television Association (CCTA). My career plan was to stay in Vancouver which we had all grown to love.

The next several years had their own ups and downs. During the time I was still President of Premier, Pay TV was introduced in Canada. One of the channels was the Playboy Channel. I was singled out on the West Coast as being the 'purveyor of porn'. In fact, the Playboy Channel was anything but pornographic – it was usually just plain silly (the picture of a naked girl riding a horse with the caption: bareback rider).

To endeavour to divert the uproar, I tried humour. At a talk to the Vancouver Board of Trade, I interspersed my discussion of the new role for television with observations that the subscribers could now decide whether they were watching bunnies on the boob tube or boobs on the bunny tube. I finally dismissed the protesters by observing that there is no fury like a woman porned.

As any reasonable person would have already figured out, this approach was a disaster. I quickly learned that you never make fun of a single issue group's single issue – they have no sense of humour about this at all. Even our house was picketed.

However we survived all that and with my other activities in British Columbia, soon overcame any problems that Premier had had with the Vancouver community.

We gradually integrated the Premier systems into the overall Rogers Cable Network and eventually changed the name of the whole organization to Rogers Cable. At an Annual General Meeting of Rogers Communications Inc., the parent company, a shareholder objected to losing names such as Canadian Cablesystems. I defended Ted by pointing out that the Rogers name had a long and illustrious history going back to Rogers Majestic and CFRB (Canada's First Rogers Batteryless), secondly the name containing the term Canadian was a disadvantage in our growing cable business in the United States and finally the only alternative was Fierheller Cable which even I would not vote for. The motion passed.

My parallel career continued unabated. The United Way of the Lower Mainland had practically helped me carry the bags off the plane when I landed in Vancouver. The United Way has a wonderful national network and Vancouver had been advised by the Ottawa United Way to look out for me.

They asked me to Chair the United Way Campaign in 1983 but the chosen Chairs for each of 1981 and 1982 were transferred by their companies back East as was not uncommon in B.C. I ended up Chairing the 1981 United Way Campaign only a year or so after I arrived in the city.

The Campaign was very successful and I became Vice Chairman of the United Way Board. I became a strong proponent of an idea that remains controversial to this day – that of allowing donors to use the facilities of the United Way to give to non-United Way Agencies. This 'donor option' approach was necessary, I felt, because without this United Way might lose its dominant position in Employee Campaigns. It has been adopted by United Ways across the country in varying degrees.

However, not everything was smooth during my tenure with the United Way of the Lower Mainland. I strongly disagreed with an attempt by the staff to integrate the local Social Planning Council with the United Way. While I had no problem with the United Way funding an independent planning council, I felt that the United Way should not become a broad based social advocacy group as this could interfere with its primary fundraising role. The somewhat left wing staff led a revolt against my becoming Chair of the Board. I took the position that the choice of Chair had to be made by the volunteer board members, not by the staff. The Board vindicated my stand but given this and my problems with the Playboy Channel, I declined to serve as Chair.

The whole controversy stirred up great media interest. I remember being interviewed by Ben Webster, the controversial radio commentator, and jokingly pointed out to him that at least I had been vindicated by a vote of 17 to 8 with 2 abstentions. I noted that he had not likely received that kind of endorsement! Ben and I became good friends and he was the MC at a farewell dinner held for me at the Pan Pacific Hotel shortly after I elected to return to Toronto.

With my continuing interest in higher education, I agreed to serve on the Advisory Board of the UBC School of Business. However it was Simon Fraser University that first approached me to serve on their Board. I did so and Chaired their Finance Committee until I finally left Vancouver.

I found the time, or as Glenna said, I made the time, to serve on the Board of Vancouver Opera, the Vancouver Centennial Commission, the Vancouver Chamber Choir and was on the Founding Committee of the new Vancouver General Hospital Foundation.

When B.C. went into a slump in the early 80's, I was asked by the B.C. Government, which was Social Credit at the time, to set up a public/private organization to try to revitalize the economy. This came to be known as Team B.C. which worked primarily with smaller communities. For example, if a community was dependent on the logging industry and this was lagging, we tried to convert them to tourism or some other pursuit. Grace McCarthy, the Deputy Premier was very supportive of this effort and it became something of a model for other regions of Canada experiencing economic difficulties.

Vicki had moved to Vancouver after completing her course and started to work with a prominent veterinary clinic. Lori, after her graduation with a B.Sc. in Geological Science, was having trouble finding work in her field. There were not many women Geologists as they actually proved to be a nuisance on field parties as they required separate facilities. Lori therefore started to work at the same veterinary clinic.

After a while they learned that the owner had a run-down animal boarding business in Richmond, B.C. just south of Vancouver. He was thinking of selling it but Vicki and Lori persuaded him to put some money into it and they would run it.

The venture was quite successful. It was a large operation with facilities for 100 dogs and 25 cats. They actually moved to a mobile house on the property.

In the meantime, I was trying to lever the money that I had accumulated from the sale of SDL. At that time the real estate market on the West Coast was booming and I invested heavily in properties in B.C. and even several in Hawaii. When purchasing the investments, I might have missed the top of the market by a couple of days but not likely more than that. The real estate market went into a slump and interest rates escalated well into double digits. Much of the capital I had built up disappeared. It was obvious I was going to have to start over again but as usual my next career move was no better planned than any of my prior moves. It just happened.

Now We 'Cantel' the Story:

I was about to move from the Wired World of cable to the Wireless World.

At the time I first heard about cellular telephones, I was still the President of Rogers Cable Systems on the West Coast. My only exposure to the telephone was as a user, who like most people, did not even know how to use most of the features on my office phone.

One day in 1982 the owner of a small Vancouver paging company, King Margolese, approached me to talk about how a cable company might work with his organization which wanted to bid for cellular telephone licenses. I sent a memo to Ted Rogers and put together the best information I could find about this new industry.

Ted knew as little about the field as I did but with his keenly developed sense of what could be a new field with a lot of potential, Ted moved quickly. In the meantime, Marc Belzberg, Sam Belzberg's son, was in New York researching firms in the United States that might be good investments. He noted that some of the rising stock performers had an interest in the cellular industry. Ted, Sam and I had had dealings in the past and the contact was quickly made.

At the same time, Philippe de Gaspé Beaubien of Telemedia, my director friend from the SDL days, had a Vice President, Corporate Development, who was looking for new opportunities. David Lint quickly turned up the cellular industry as an interesting possible venture. Philippe was also a good friend of Ted's.

This group quickly evolved into a consortium to bid for the cellular licenses in a number of cities across Canada.

As Ted light-heartedly noted, this all-Canadian consortium had all the ingredients necessary – a Jewish family from Western Canada, a Protestant family from Ontario, and a French Canadian Catholic family from Quebec. It was a prophetic observation.

The consortium shareholders asked me if I would assist in steering the application through the bidding process – just a few weeks work. Having been through a start-up situation of a new company, I should have known better. It was going to be a long few weeks.

We started in late '82 with just a 'pick-up' team of people we had borrowed from the founding companies or brought in under contract.

Charles Dalfen of Johnston and Buchan, a well known Ottawa legal firm, assisted us with the regulatory preparation. He later became the Chair of the CRTC.

David Lint was seconded from Telemedia to work on the marketing and other strategic areas as our Vice President, Development.

Pierre Morrissette, who brought a strong financial background from Telemedia, was loaned to us as Executive Vice President.

King Margolese became our Vice President, Radio Common Carrier Operations.

Nick Hamilton-Piercy was loaned to us from Rogers as our Vice President, Engineering.

Finally, I was appointed President or I suppose any other title I wanted to take at that time. The positions were all volunteer and unpaid, at least by Cantel.

We set up a war room in the Canada Consulting Group offices above a liquor store on Front Street next to the St. Lawrence Market. It was to become our home away from home for the next several months.

We hastily designed an interim logo and proceed to submit an Application "on behalf of a company to be incorporated". In fact, the actual Memorandum of Understanding amongst the three Partners was not signed until February 25th, 1983.

At this point we made a crucial decision. Not only would we submit an extensive volume (about 1.5 inches thick) for each of the 17 areas, but would duplicate this in French. The logistics were a major undertaking as we also decided to submit

Applications simultaneously in each of the Department of Communication (DOC) Regions. The Regional Directors were going to participate in the decision and we felt they deserved the direct attention. Brent Belzberg and I delivered a set of Applications in Vancouver on February 28th. The other members of the team flew to other parts of the country with boxes of Application sets.

We had suggested to DOC that personal presentations be allowed and in a letter from DOC on June 20th, 1983, the Applicants were requested to make a presentation during the first week of July. The presentations were held in the Ottawa Conference Centre. Fittingly, the table around which we sat was in the shape of a cell.

In what was already becoming Cantel style, we decided to make a major audio/visual presentation. When the day arrived we had about 25 people in the room including the three Principals – Philippe, Marc and Ted.

Our single 'pitch' was that Cantel's superiority stemmed from five factors:

- The all-Canadian ownership.
- The substantial financial operational resources.
- The consumer service expertise.
- The technology management capability
- The appropriate structure we were proposing.

We also undertook to use Canadian technology wherever we could. To this end we had been working closely with NovAtel of Calgary.

Included in our presentation was a comparative analysis of Applications which needless to say pointed out that Cantel was the only logical candidate. This raised some eyebrows at the DOC who subsequently asked that a chart be distributed to the other bidders for comment.

The original concept was for the Cantel Cellular Radio Group Inc., as the newly incorporated company was called, to be an umbrella organization with separate operations in the West where the Belzberg Group would have the majority control, in Ontario where the Rogers Group would be the dominant factor and in Quebec where the Telemedia Group would take the lead. It was an aggressive presentation.

On August 5th, I wrote a memo to the Board indicating that I felt a decision had already been made. As I pointed out, my salesman's gut feeling was that we were the leading contender. However, much more was going to happen before a final decision was made. On August 18th, the then Minister of Communications, Francis Fox, advised that the DOC had made the decision that Applications would only be considered for all the 23 Metropolitan Areas. This was a crucial and very wise decision as it meant that only one company would be licensed to compete against local telephone companies across Canada. This was quite different from the approach taken in the United States but one much better suited to the Canadian environment. Amended proposals were to be submitted by October 14th.

On December 14th, 1983, I received a call from Francis Fox to advise us that Cantel had been awarded the National Cellular Radio Licenses. The excitement was only tempered by the realization that we now had a huge job ahead of us. Peter Newman called this win "1983's most impressive business coup" (Maclean's, January 2, 1984).

That evening Philippe called me in Vancouver and said, "For heaven's sake, George, what do we do now?"

The Shareholders asked me if I would stay on to help get the group organized. I could see it was going to be a busy Christmas Season.

Over the next couple of weeks I started by listing down everything I could conceive of that would need to be done to get an organization in place, arrange financing, create a new image for this unknown company, hire staff, design the system, implement a marketing strategy, and find space from which we could operate. I then divided the tasks up as best I could amongst the group who had worked on the Application.

On January 4th, 1984, I submitted a rough budget for what I thought it would take to get us through the initial stages. The amount was $1.5 million.

The Partners were shocked! I believe they had really thought that the operation would be run out of offices already existing in, say, Rogers Cable or Telemedia. I had a far different vision which was to capitalize on this incredible award. For the first time, Canada had the opportunity to have a national coast-to-coast Canadian telephone company.

Things were not to run smoothly, however. Almost immediately Bell Canada announced its intention to start providing cellular service in Montreal and Toronto in September, 1984. I immediately telexed Francis Fox to indicate that this was a totally unfair procedure. In our submission we had never implied it would be possible for the non-wireline to start before May 1st, 1985, at the earliest. My arguments against this head start were accepted and in a crucial ruling on March 14th, 1984, Mr. Fox announced that neither company would be able to start until July, 1985.

As it was absolutely essential that Cantel have the right to inter-connect with each local telephone company, it was further decreed that elsewhere in the country a telephone company would only be able to start service six months after an inter-connect agreement had been completed with Cantel.

These two key decisions were to play a major role in the phenomenally successful launch of cellular in Canada. The fact that both parties started simultaneously with all of the market hype meant a much faster start than had been the case in the United States. There the wire line companies were allowed to start when ready and were only required to provide reseller services to the non-wire line company.

Shortly thereafter, the Canadian Radio-Television and Telecommunications Commission (CRTC) issued a major set of rules and regulations governing the cellular radio development in Canada. This decision, CRTC 84-10, amongst other things, confirmed that Cantel would be a telephone company operating under The Railway Act. It further defined the method of operation and allowed Cantel the right in Federally regulated areas to carry its own long distance traffic. This was later to prove crucial as this additional revenue would help Cantel expand to less densely populated areas.

Having now gained the necessary time to do a proper planning job, we moved ahead quickly with staffing up the organization.

We hired Walter Steel as our new President. Walter had been the founder of AES Data Systems and we felt, therefore, had the know-how to guide the Company through a start-up situation.

Walter, in turn, hired Paul Kavanagh as Vice President, Marketing. Joe Church was hired from Bell Canada as our Vice President, Corporate Planning and Development. David Perks was brought in as our new Vice President, Finance, and Marc Ferland joined us as the Vice President of our Eastern Region.

Two other key players joined Cantel at this time. Nick Kauser was the only one of the group except for Joe Church who had experience in a telephone company. Nick had run his own company in Venezuela in the telephone field. He would prove to be the driving force in getting this new telephone company built in less than 18 months. When Ted Rogers interviewed Nick, he asked him if this was possible. Nick said simply, "Do we have any choice?"

Roger Keay was the only member of the original team, other than myself, who now moved over permanently with Cantel. He reported to Nick.

The other key player was David Parkes who was hired as the Vice President, Central Region, which was Ontario less Ottawa.

The team was coming together. Many of us were from outside Toronto and were living at the King Edward Hotel – soon to be known as Chateau Cantel.

However, this new team needed an office. When I started setting up the operations in January, 1983, I was borrowing space wherever I could find it. Largely this was at the Rogers offices in the Commercial Union Tower in Toronto. I literally used anyone's desk who was away for the day.

The entire set of corporate files was in an over-sized legal briefcase that I carried with me wherever I went. With the help of Barry Ross, a Senior Vice President of Rogers, we located our initial space at 20 Queen Street West in the Cadillac Fairview Tower.

I had already hired Julie Robson as my Administrative Assistant. Julie doubled as Walter Steel's secretary and rapidly developed a reputation for bringing this lively group back to earth. At one point, we were having trouble finding a suitably high cell site in Montreal. Julie observed that if we piled our egos one on top of the other, we should be able to reach about any height we wanted!

There is more behind this observation than one might realize. The start-up group at Cantel had taken on the momentous job of competing with Bell Canada by announcing it would start simultaneously in Toronto, Montreal, Hamilton and Oshawa on July 1st, 1985. At this point most of us still knew little about running a telephone company and none of us accurately foresaw the incredible challenge ahead of us. This may have been an advantage as we just did things not knowing they should have been impossible.

In Montreal we had selected premises on Cote de Liesse which would be our Head Office although our Executive Offices were going to be in Toronto. This put executives in each of the two major centres we planned to serve initially.

I likened Cantel at this point to a ship with three prows as each of the founding Partners had a slightly different perception of what they expected from the Company. The Belzbergs believed the Company should be run frugally and should concentrate on building in major areas providing the fastest return.

Philippe was more of a visionary but headed what, at that time, was a relatively small company that would have had difficulty keeping up with the financial requirements for a major expansion. Ted was also a visionary whose view of the opportunity was starting to change. While initially he felt the Company would simply provide service in the minimum number of areas, he was starting to see the incredible potential for a national telephone company.

It should be noted that the Board of Rogers Communications Inc., which was the publicly traded company, had elected not to invest in Cantel. Ted took the investment in Rogers Telecommunications Limited which was his private holding company. There was, however, an agreement that the shares of Cantel could be taken over by the public company should the Board of that company so wish. RCI at that time was heavily extended in Cable in the United States and could not afford to finance a huge expansion such as that already envisaged by the senior Cantel management group.

"Get me outta this Board (bored) Meeting!"

With these diverse interests and capabilities, it was not easy to get a single direction for the Company. An Executive Committee was established under the direction of Harold Nickerson of Telemedia. Robi Blumenstein of First City served on this Committee and I represented the Rogers group. Even so, simple decisions were often hard to come by. As a result, Cantel management developed a style of simply doing whatever it had to do to get the job done.

Some very difficult decisions had to be made. The first was the development of a distribution approach. Walter, Paul Kavanagh, and the two Regional Vice Presidents came up with a strategy of establishing Cantel Service Centres. These would be independently owned entrepreneurial operations serving particular geographic areas. They, in turn, could recruit whatever Agents they wished. Rather than establishing our own Sales outlets, this allowed us to start a number of areas in parallel. The approach proved to be one of the keys to our success.

A second major decision was, of course, the choice of network equipment. We had done all we could to make use of the Canadian equipment being designed by NovAtel. However, it became clear that there was no way this equipment would be ready in time for a July 1st, 1985, start. We therefore made the decision to use Ericsson equipment. This turned out to be an excellent choice although it required some selling to the Department of Communications and a number of commitments by Ericsson to develop an R&D Centre in Canada amongst other things.

In many ways, Walter and his team had done an outstanding job in launching a number of these concepts.

However, keeping Walter and the Board in step was just as big a problem. At one Board Meeting held at the Royal Canadian Yacht Club in the Fall of 1984, Walter chose to lecture the shareholders on the wonders of matrix management. It was clear that he was not enjoying the confidence of the Board. Marc remarked that he could not see the reason for anything so complex. He said that you just told people what to do and if they didn't do it, you fired them. Ted looked at the matrix and commented, "It looks like communism to me".

The situation went from bad to worse with an apparent lack of communication between the President and the Board. Finally it fell on Harold Nickerson, as Chairman of the Executive Committee, and me as Chairman of the Board, to meet with Walter and come to a parting of the ways. The meeting took place on neutral ground in the offices of Tory, Tory, DesLauriers & Binnington.

The partners were still having difficulty coming to grips with the financing for the ever-expanding capital needs of Cantel. By the Fall of 1984 Cantel had made a couple of attempts at a private placement in Canada. Neither of these were successful. Finally, Marc Belzberg made contact with Ameritech Mobile Communications Inc. of Chicago. Ameritech had been the pioneer in cellular in North America, conducting the first experiment in the early 80's with about 2,000 cellular phones. It was felt that this group could bring some helpful experience to the emerging Cantel organization as well as some much desired funding.

On December 14th, 1984, the Partners obtained the concurrence of the then Minister of Communications, Marcel Masse, to allow Ameritech to take a 19.99% interest in Cantel through a Canadian subsidiary called Pan Canadian Communications Inc. The three founding Shareholders confirmed their commitment to ensuring Cantel would have a Canadian equity base of $15 million including $9.1 million of their own equity. The rest, it was anticipated, would be raised by a public issue or private placement.

It was to be another disrupted Christmas vacation. Incredible though it may now seem looking back, it did not prove possible to raise sufficient equity on any basis from Canadian sources. The original Partners were on the hook for the total commitment.

By the end of February, 1985, it was clear that we would not meet the payroll unless additional equity was forthcoming from the founders. We had stretched our credit with our suppliers to the limit. Just before the 60 days were up, sufficient equity was put into the Company to meet our commitments.

With Walter gone, we needed a new President and July 1st was quickly approaching. I had originally tried to hire John McLennan as President of the Company but John was enjoying his new role as a Consultant after a number of years in a senior position with Mitel. John did, however, come on our Board. At this point Harold and I approached him to see if he would take on the job as President for at least an interim period until we could finish the search for a new permanent President. John agreed and took over immediately from Walter Steel.

By this time I had hired Brian Josling as our Vice President and General Manager for Western Canada and the team was pulling together in an outstanding fashion toward what must have seemed an impossible goal of being operational within the next few months.

The Company had now outgrown its premises at 20 Queen Street and had moved into a new building at 40 Eglinton Avenue East.

The group now met regularly to finalize the plans for our launch. Julie Robson kept the Minutes of these meetings and continued to bring to earth this group of high-fliers. In one meeting, Nick Kauser asked Julie if there was any coffee. Julie replied that she didn't know but if Nick found any she would love a cup.

It was clear that Bell Canada did not really take us seriously at this point. They could hardly credit that within a few months a group starting from scratch could have put together not only an effective organization but put up a cellular system that proved to be superior in terms of coverage and performance and then out-market Bell.

The opening celebrations were well planned. At the pre-launch in Montreal Philippe de Gaspé Beaubien made the first official telephone call on the new network. Naturally, Philippe knew nothing about how to operate a cellular phone and asked Nick Kauser for assistance. Nick suddenly realized that he had never made a cellular telephone call either but remarkably the first call went through with no problem.

On the launch day, we had arranged a call from Mayor Art Eggleton from Nathan Phillips Square to Mayor Jean Drapeau who, at that time, was visiting the Ramses II Exhibition in Montreal. I announced with great pride that this was not only the first official cellular telephone call but the first known telephone call to an Egyptian Tomb!

By August 31st, 1985, which was the end of our Fiscal Year and two months into operation, we had over 3,000 subscribers and were clearly on our way.

However, we still did not have a permanent President. John had let it be known that he did not want to stay on beyond August 31st despite our pleadings. John had done an outstanding job during this interregnum.

As Chairman, I had launched a search for a new President. However, Denny Streigl, of Ameritech, then Chairman of the Executive Committee, called me in early August to ask if I would return to the President's job, give up my Cable interest in Vancouver, and move to Toronto.

For the past couple of years, I had commuted from Vancouver to Toronto almost every week. I was really not living in either city but as I jokingly said, spent most of my time, 30,000 feet over Winnipeg.

On September 1st, 1985, John McLennan and I changed jobs. At one of the periodic Fireside Chats that I customarily held with staff, I had the amusing job, as the Chairman, of congratulating myself as the incoming President.

As might be expected, there had been a considerable let-down following the July 1st launch. We brought up Ottawa and Quebec City on schedule and had planned a major roll-out across the rest of the country. However, it was becoming clear that the organization needed some changes if it was to sustain itself during the next growth period.

To get the maximum sales activity underway simultaneously, we had decentralized the operation quite substantially to the Regions. As in most companies, this brings with it some problems as well as some potential. The problem was that we did not have the policies in place to ensure continuity of service in the various Regions. It became necessary, therefore, to centralize the control of our operations to a much greater degree until such time as we were ready once again to delegate back to much strengthened Regions.

The second difficulty surrounded the splitting of our operations between Montreal and Toronto. This simply did not work. We gradually expanded the Executive Offices in Toronto. This left Joe Church as the only Head Office Executive in Montreal. As Joe did not want to relocate to Toronto, we also came to a separation agreement.

However, not all the problems of the multi-prowed ship were cured. It was becoming more and more obvious that one single Shareholder needed to take responsibility for the Company. Ameritech could not as they were limited to less than 20% of the equity. First City was not interested in assuming a larger role in the direction of the Company. This left it to either Rogers Telecommunications Limited or Telemedia to assume the dominant role. At a meeting held at Tory's, both Ted and Philippe put forward their case as to why their organization would make a good lead Shareholder. By the end of the meeting it was agreed that Rogers Communications Inc. would acquire a 33% equity interest in the Company with a 63% voting interest. First City and Telemedia would each reduce their interest to 11.5% of the equity and 3.2% voting, and RTL, the original Rogers Shareholder, would maintain a 10.6% interest and the Radio Common Carriers, who had been provided with an equity opportunity in the Company, made up the difference.

In May, 1986, Ted Rogers assumed effective control of the Company and became its Chairman.

In the meantime, we had opened in Vancouver in January of 1986 to be followed by Edmonton, and Calgary. The later two centres were our first experience in dealing with Provincially regulated or Municipally regulated telephone companies. We were to become very used to the intricacies of negotiating with Public Utilities Boards across the country to try to get suitable inter-connect agreements in the areas in which the CRTC did not have jurisdiction.

To assist in this, we brought in Barry Singer as our Legal and Regulatory Counsel. Barry had a difficult first week at Cantel. At his initial Operations Committee Meeting, we asked if there were legal problems he had found. He said only one which involved a possible wrongful dismissal action. However, he noted that fortunately it was just a secretary. There was total silence around the table as the management group looked warily at Julie. Without raising her eyes from her note-taking, she simply commented, "Then just any lawyer can handle the situation".

Like many senior executives, I always disliked doing appraisals. Telling people what was wrong with them is not exactly my style. Realizing that I had to do this for my Executive Assistant, Julie Robson, I approached her and said "Julie, it is time for our annual appraisal." "Fine" she said, "Take me to lunch and I will tell you how you are doing."

Julie got headaches from red wine but loved Champagne. I arranged for a limo to take us to lunch and had a ½ bottle of Champagne in the back seat. As I recall, the appraisal went just fine although I cannot remember too many of the details.

In the meantime, sales were expanding faster than anyone anticipated. For our Fiscal Year ending August 31st, 1988, we had originally forecast 60,000 total subscribers. By September 1st of 1987 we had revised that estimate to 75,000. By the end of the Fiscal Year we had exceeded 95,000.

We now had out-grown 40 Eglinton Avenue and in 1988 made the decision to move our Executive Offices to 10 York Mills Road in Toronto. By August 31st, 1989, the staff had grown to over 900 to service the subscriber base that had now grown to over 150,000.

As our network continued to expand, and we added more and more of our own microwave towers, we could also look at adding new services.

John Lang joined the Company as our Vice President and General Manager of the newly established Paging Division. The intention was that on receiving national licenses for Paging, we would use our current infrastructure to provide a new 900 MHz national Paging service. In January, 1990, we received these licenses for both national and international Paging.

Tom Pirner also joined the organization as head of our new Mobitex Data Radio Division. This new service would provide a primarily data-oriented system for trucking, dispatch, taxi or even inventory taking applications using different frequencies than those available for cellular but still using the same national network of towers and microwave.

By the end of our 1985 Fiscal Year, we were serving eight of Canada's ten Provinces and had expended nearly half a billion dollars in network capital.

Cantel had garnered substantial recognition for its marketing efforts having won the Canadian Award for Business Excellence in Marketing in 1987 and having been used as a case study for the University of Western Ontario's MBA Programme.

On the technical side, we had developed an international reputation for pioneering in the cellular field and were now being invited to bid on overseas contracts.

Cantel had an interesting history as a company sometimes with public investors and sometimes privately owned by Ted Rogers or Rogers Communications Inc. At one point we had decided to take the company public and had arranged for a Road Show for the IPO that covered about 15 cities in 9 days. To accomplish this, our underwriters had chartered a Lear 25 – not a large jet! When we finally completed this whirlwind tour in San Francisco, Bruce Day, our Vice President Finance and I were left alone to take the return non-stop flight to Toronto on the tiny jet. To celebrate this hectic but successful tour, I bought a bottle of Russian River Chardonnay and Bruce bought some beer. I proceeded to down the entire bottle of Chardonnay forgetting that the Lear 25 did not have any washroom facilities. I solved the problem by recycling the Chardonnay back into the original bottle somewhere over Chicago!

We capped off the end of our Fiscal '89 Year by announcing a $600 million expansion over the next several years to provide Canadians with the world's longest continuous coverage corridor, coast-to-coast.

In September, 1989, Ted and I had lunch at the Granite Club on a Saturday to discuss the next steps for Cantel. It was becoming obvious that the Company needed a full-time Chief Operating Officer.

Ted suggested that Jim Sward, who was then President and C.E.O. of Rogers Broadcasting, would be interested in taking on this challenging task. Jim joined us in October of 1989 as President and C.O.O. while I reverted once again to being Chairman and C.E.O.

Cantel was now a broadly based mobile communications company. By the end of 1993, Cantel served over 500,000 cellular customers alone with many more through our Paging and Mobitex Divisions.

I could not begin to acknowledge all the people who helped to launch Cantel. The first five years since we started offering service demonstrated that a dedicated, fun-loving group of young people can start a business against major entrenched competitors and succeed.

I stayed on as Chairman & C.E.O. until 1993. Jim Sward had resigned at that time and we brought in a new President from the United States, David Gergacz. He also took over the title of C.E.O. and after a decade I stepped aside from active management of the company.

I was given the title of Honorary Chair of what is now Rogers Wireless Inc.

The company has continued to prosper and by 2004 had revenues of over $3 billion, 5,500,000 subscribers and 3,000 employees.

My personal capital had rebounded as a result of shares and options in Rogers Wireless. I then moved to the RCI Head Office as one of two Vice Chairs reporting to Ted Rogers, the other being Phil Lind.

My major responsibility was to be Roger's 'Ambassador' as someone described it. I undertook a number of industry association positions representing Rogers such as Chairing the Information Technology Association of Canada in 1993-4, serving on the Board of CANARIE (the developer of the Canadian Internet Network), participating in the founding of Smart Toronto to promote high tech industries in the Greater Toronto Area and similar industry functions.

Ted Rogers had always had an ambition to spread his media empire abroad. During my time as Vice Chairman, RCI, I had led a team to try to win a cellular telephone licence in the newly established Czech Republic, immediately following the Velvet Revolution. In fact we were in Prague only about three weeks after that event.

When we met with government officials, we found that the Minister of Communications had been a school teacher a few weeks before. The government was more interested in how to run an election (we arranged for the Chief Electoral Officer of Canada to send someone to help them out) or how to de-bug their buildings. The Communists had placed bugs everywhere and they had a concern that these were still being used to monitor their activities (we arranged for equipment to be sent from New York).

We did not win the business despite having a high powered consortium composed of British Telecom, Nokia, McCaw Communications and others. We were outbid by Bell South. However, it was a fascinating experience.

Ted then asked me if I would do a tour of the Far East to investigate possible opportunities for business. There were certainly impressive opportunities, the best of which appeared to me to be a joint venture with NT&T to open up cable television in Tokyo. However it was very obvious that Ted did not have the enthusiastic backing of other members of management. I recommended therefore that rather than trying to do something on his own, he become a partner with Charles Sirois in the Télésystem International Wireless venture which he subsequently did.

To show the far ranging interests that Ted had, while I was in Hong Kong he asked if I would take a quick trip to Macau to check out casino operations. Ontario was about to grant licenses for gambling in the province. I soon found that the gambling operation in Macau was ruled by Stanley Ho. He not only owned most of the casinos but had a controlling interest in many of the hotels and even owned the jet boat line from Hong Kong. I met with him briefly. One of his Lieutenants gave me a detailed tour of their back room operations. While it was fascinating and the potential was obviously huge, I recommended that Ted not get involved in Ontario. My concern was that of image even though the profits looked tempting.

I always seem to get in trouble in Paris but on reflection realize that I had just as many difficulties in London. On returning from a Falconbridge Board Meeting in Norway, I planned to stay over for a night or so in London. As usual I had over planned and had arranged for dinner at Le Gavroche. This has always been one of my favourite London restaurants being run at the time by Michel Roux and his brother. Michel now runs the Waterside Inn, another Michelin starred restaurant. I arrived at Heathrow mid day and took a taxi to a small hotel I sometimes used called the Halkin (about a block south of the Lansborough at about half the price). About half way in from the airport, the taxi started to slow down and finally broke down entirely. We had to flag down another taxi and the driver of the first cab asked if he could get a lift in town and then further asked if we would not mind dropping him off at his house. I must have been very accommodating but we managed all this. I finally arrived at the Halkin in mid afternoon.

I had wanted to see the new Saatchi Gallery in North Central London at St. John's Wood and immediately grabbed another cab there. Naturally, I had not checked that the Gallery was closed on Thursdays. Another cab ride took me to the British Museum where I had wanted to see some Illuminated Manuscripts in any case.

I finally arrived back at the hotel but was still too early for dinner. I wandered down the street with some reading material to what appeared to be a lovely little enclave called Belgrave Park. I pushed open the gate and spent a pleasant hour reading in the park.

However, when I tried to get out of the park all the gates were locked and the fence was about 6 feet high. I realized that I was in a private park. Fortunately, a young couple in tennis whites were coming into the park for a game and opened the gate for me. "Oh did you forget your key sir?" I mumbled something about having Alzheimer's and left them with a cheery "tallyho" and got back to the hotel just in time to catch another cab to Le Gavroche. That also turned out to be something of a disaster as I was presented with a quail egg appetizer which I assumed was hard boiled. I bit into it and found that it was anything but. I squirted egg juice all over the menu, the head waiter and I suspect the diners at the next table. The waiter was very solicitous and said that "no problem sir this happens all the time". I suspect he was really thinking "dumb colonial".

In any case this was a typical day in London and the reason my wife only travels with me periodically!

When we did travel to London together, we tended to stay in Room 610 in the Savoy. This was the room in which Monet stayed and painted his famous scenes of the London Embankment. I had made reservations at Chez Nico, at that time the only Michelin Three Star restaurant in London but when we tried to get to our dinner along the Strand, we found it was as usual under construction. After a number of frustrating delays, the typical London cab driver looked back and said "don't worry Guv, I'll get you there in time for breakfast!"

Another typically disastrous day in London occurred when I was returning from a Telesystem International Wireless Board Meeting in Basingstoke. I was staying at the Mandarin Hyde Park and as usual had arranged for a dinner in London before returning. This time I selected Pétrus. This was a newly opened restaurant and I had asked for a reservation at 7:30 p.m. on a Saturday evening. The restaurant called the Concierge at the Hyde Park and asked if I could accommodate them by coming at 7:00 p.m. so they could do two sittings. I agreed but asked the Concierge to ensure they provided me with a complimentary glass of Dom Perignon for being so helpful.

In the meantime I had booked a matinee to see 'Copenhagen' which I came to regard as one of the outstanding plays of recent years. After the performance I had time to kill before 7:00 p.m. and walked back from the theatre. The restaurant was then on St. James Street (it is now next to the Berkeley Hotel) and I found myself there at about 6:00 p.m. To fill in the time I decided I would have a drink at the Ritz. I had not noticed that it was their Saturday Happy Hour and the Martini I ordered was the size of a small bathtub. After downing this, I proceeded to the restaurant where they indeed produced the glass of Dom Perignon. Being at Pétrus (although it has no connection to the vineyard) I then ordered a bottle of 1982 Pétrus at a cost just slightly under the GDP of the United Kingdom.

After a delightful dinner I must have shown some signs of inebriation as the waiter politely asked "sir would you like us to call a taxi or an ambulance?" There was always something about the British sense of humour that appealed to me.

I finally retired from any active management position at Rogers Communications Inc. on December 31st 1996 although as described later, my retirement is treated as something of a family joke.

I took an office in the core of the city and am now located on the top floor of the Royal Trust Tower in the Toronto-Dominion Centre on King Street together with a number of other 'used executives'. Together we share boardrooms, fax machines and administrative assistants. It works very well as it keeps me out of Glenna's hair as well as keeping me close to my other activities.

My parallel career had continued during my return to Toronto. In 1991 I Chaired the United Way Campaign and then became Chair of their Board. This was becoming something of a habit as I had now Chaired United Way Campaigns in each of Ottawa in 1971, Vancouver in 1981 and Toronto in 1991. I claimed that I would only consider a United Way Campaign in 2001 if it was in Tahiti!

I was also Chair of the Toronto Board of Trade from 1996-97. More correctly, I was the last volunteer President. For about 150 years, the Toronto Board had always referred to the top paid executive as the General Manager. I felt that this individual should be the President and C.E.O. and the top volunteer should move to being Chair. Elyse Allen became the first staff President during my term. The Board, which was one of the largest in North America with over 10,000 members, became even more proactive in the policy field during her term.

My retirement now allowed me to convert my parallel career into a full-time job – that is where you work just as hard as you always did but no longer get paid for it!

I was asked by the Association of Collegiate Entrepreneurs (ACE) to give a talk in Montreal on career development. As pointed out elsewhere, this was hardly an area of expertise for me. Besides making the usual comments about it being helpful to have been born at the right time e.g. in the 1930's and generally be in the right place at the right time to take advantage of any new trends, the only really useful advice I felt I could pass on was my experience with my parallel career.

I pointed out that the experience one gains in dealing with volunteers is extraordinary. Volunteers can walk away from anything you are trying to get them to do at any time. You very quickly therefore learn to motivate people to want to do whatever you need them to do. Inevitably this is better than ordering them to do something that they are baulking at.

A parallel career can also expose you to a wide range of experiences that one might not get just in one's professional career. You meet very interesting people and learn to look at situations in very different ways.

While one does not do things in the voluntary sector for the sake of one's career, it does not hurt to be actively involved in industry associations or other professional groups. To be very crass, it is more difficult to fire someone who is the head of a national industry association than it would be to fire someone who simply does their 9:00 to 5:00 job.

I also pointed out that a parallel career can give someone a sense of confidence and self worth. Even if your professional career is not working as well as you would hope, you can always point to a series of other accomplishments.

I called this talk about a parallel career "The Competitive Edge".

My Board of Trade experience led me to the conclusion that while the Board did a wonderful job in pointing out all the things that could be improved in the City, they did not see it as part of their mandate to promote the area as a great place in which to live and invest.

Mayor Hazel McCallion of Mississauga had established an informal committee composed of the Mayors and Regional Chairs of the 29 municipalities in the Greater Toronto Area. This group became a proponent of establishing an organization to promote investment in what is now the 4th largest city region in Canada and the U.S.

With the help of the Toronto Board and the Boards of Trade and Chambers of Commerce in the GTA, we put together an organization called the Greater Toronto Marketing Alliance in 1996. I was the founding Chair and remained in that position until 2002. We recruited Karen Campbell who had been the Chief Development Officer for Mississauga as the President. I am currently the Chair Emeritus of the organization which continues to successfully market the GTA internationally.

In a somewhat parallel area I also served for 6 years on the Board of Ontario Exports Inc. This is a Provincial Government body to promote exports from Ontario.

In the meantime I was asked to sit on the Selection Committee to pick a Campaign Chair for Trinity College, part of the huge University of Toronto Campaign. I missed a couple of meetings and this turned out to be a mistake as the Committee then asked me to Chair the Spirit of Leadership Campaign.

With the help of Kara Spence, Vice President Development, we successfully raised over $18,000,000, some of which was used to fund a new College Library called the John W. Graham Library.

As mentioned earlier, Ted Rogers and I are Sigma Chi's and had known each other since university days. His step-father was John Graham who coincidentally was also Sigma Chi and after whom the Library was named. John was a true gentleman with a delightful sense of humour and had been my most influential mentor for many years.

I was also one of the founders and subsequent Chair of the Sigma Chi Canadian Foundation that was established to promote scholarship and provide other assistance to Sigma Chi undergraduates. Needless to say one of the scholarship programs established under my Chairmanship was named for John Graham.

I had some hospital experience in both Ottawa and Vancouver and had served on the Sunnybrook Health Sciences Centre Foundation Board for a number of years. In 2000, I was asked to Chair a major Capital Campaign for about $100,000,000. I agreed but this time had not done my homework.

The Hospital which is one of the largest in Canada had just gone through a huge merger with Women's College Hospital and the Orthopaedic and Arthritic Institute. Their Foundations were still separate and it became obvious to me that they were in no position to launch such a large undertaking.

I delayed the start of the Campaign until we could integrate the three Foundations, hire a new President, integrate the staff and give the Hospital time to re-organize its needs. As it turned out the Hospital also decided to look for a new President.

The Campaign is now being launched although typically the goal has expanded to $300,000,000. This is hardly surprising given Sunnybrook & Women's role as one of Canada's largest Teaching, Research and Advanced Patient Care facilities.

In the meantime Kara Spence who had been at Trinity College had moved on to the Canadian Institute for Advanced Research. This is an organization that promotes research involving Canadians as part of internationally selected teams to examine leading edge questions in such advanced fields as Cosmology, Earth Evolution, Quantum Information, Nanotechnology, Population Health and similar fields. Their new President Chaviva Hosek and their Chair, Tom Kierans asked me to join the Board and once again I agreed as it fit very well with my lifelong interest in and curiosity about leading edge research. I now Chair the Advancement & Communications Committee.

I was also President of the National Club, 1998-9, and had the fun of taking a 125 year old establishment club and turning it into a dynamic part of the Toronto scene. The Club had been in decline in terms of membership for some years and had been losing money. With the help of the newly appointed General Manager, Bill Morari, we turned the Club around completely. We redesigned everything from the menus to the wine lists, the staff uniforms, events and just about everything else. The Club made a profit for the first time in a decade and we added nearly a hundred new members.

Noting that it was the National Club's 125th Anniversary, the Club having been founded in 1874, I wrote a short, not too serious, history. We then had a celebratory dinner with such diverse speakers as Lieutenant Governor Hilary Weston and Dave Broadfoot.

I was also asked to join an organization called the Toronto Adventurers' Club. This is a group of 25 investors who meet monthly to explore investment opportunities. Their methods were somewhat haphazard but the results were quite remarkable financially. I Chaired the group in my second year as a member in 2002-3. As the organization was just coming up to its 25th Anniversary, I wrote a tongue-in-cheek history for the occasion. I suppose 'historians' often end up writing about events they did not personally experience.

I have been fortunate to work with many women executives. I have found them not only imaginative but able to get things accomplished with a minimum of rancor. Over the years I have tried to promote women to executive positions wherever I could. Kathy McLaughlin became our Vice President, Marketing at Cantel. Janice Moyer was the President of ITAC when I was the Chair. Anne Golden headed the United Way. I promoted Elyse Allen to President of the Toronto Board of Trade.

Karen Campbell became the President of the Greater Toronto Marketing Alliance while I was Chair. Kara Spence was the Executive Director for my Trinity College Campaign. Dianne Woodruff followed me as the first woman President of the National Club. Chaviva Hosek is the President of the Canadian Institute for Advanced Research just to mention a few.

In fact, other than my male Saluki and male cat (who is neutered and so is somewhat in limbo), I seem to have been surrounded by women. At one point one coyly asked if all the women I worked with fell in love with me. "Absolutely not" I replied. Then on reflection added "most do, but definitely not all". This led to some well deserved negative comments about bloated male egos!

As would be no surprise, I am also on the Campaign Cabinet for the new Four Seasons Centre for the Performing Arts – the world class facility being built to house the Canadian Opera Company and the National Ballet. This will be the only true Opera House in Canada.

I am also on the Board of the Council for Business & the Arts in Canada.

I had been selected in 2001 by the Association of Fundraising Professionals to be the Volunteer of the Year during the International Year of the Volunteer as declared by the United Nations. I was asked to say a few words at the ceremony and amongst other observations pointed out that I seemed to have spent my first 40 years building up capital and the next 4 years giving it away. I had to admit that I could not really see much logic in this but I could not ask anyone else to do what I would not do. Giving money away simply comes as part of fundraising.

Small wonder that my friends cross the street when they see me coming!

For these various purposes over the years, I had to have portrait photos done. One Toronto based photographer was clearly having difficulty with a subject with thick glasses and thin hair. After he had fussed for some time with the lights and cameras I commented that "Karsh had never had any trouble with me". The photographer said nothing but continued to work. When he finally got the arrangement he wanted, he asked when I'd had my portrait done by Karsh. I replied that I never had. I had just noted that Karsh had never had any trouble with me. I thought he was going to throw the camera at me!

I had always done some outside commercial board work in fields as diverse as Morrison Lamonthe, a major baker in Ottawa, to Falconbridge, which is one of Canada's largest mining companies.

One of my more interesting commercial board experiences was with Charles Sirois. Charles had started his career in Quebec with a small paging company. He was actually an Agent for Cantel shortly after we started in the mid 80's. We had become quite good friends.

His company was subsequently acquired by Bell Canada and he levered the proceeds from this to purchase control of Teleglobe, Canada's international telephone carrier. I joined the Teleglobe Board at this time and we saw the organization go through a great period of growth.

He then asked me to serve on the Board of a new international cellular venture called Télésystem International Wireless. TIW expanded into cellular ventures in China, India, Mexico, South America and Europe. It has now consolidated its operations largely in Europe and remains a very successful venture.

Teleglobe was less fortunate. I left the Board at the time of a merger with a U.S. based telephone firm. Subsequently, Teleglobe was bought over entirely by Bell Canada and ended up being one of the largest write-offs that Bell ever had.

However, when you reach age 70, you have to retire from many commercial boards. I had been on the Board of a Montreal based investment company, now called GBC North America Growth Fund Inc., for over 30 years. It had been one of the original investors in SDL. I had to leave this and some other commercial boards for that reason.

I remained on the Board of Rogers Wireless Inc. as their Honorary Chair until 2005 and still serve on the Board of Extendicare Inc., one of North America's largest Nursing Home enterprises.

You can stay on voluntary boards for as long as you can walk, talk and donate but this does not always apply to commercial boards.

Back to the Family:

When I had agreed to return as President and C.E.O. of Cantel in August 1985, it gradually became obvious that I could not run the company from Vancouver. As was usual, I moved to Toronto in advance of the family taking a temporary executive suite on Bay Street. I then proceeded to look for a house. The Toronto housing market was so hot at the time that people were actually advertising that they were looking for a house in a particular area.

I happened to be driving by one location when they were just putting the sign on the lawn. I came, I saw and I bought! Fortunately it worked out extremely well and we have been living happily at 24 Pearwood Crescent in the Bayview/York Mills area of the City of Toronto since 1985.

Vicki and Lori had decided to stay in Vancouver running their pet boarding operation. However, after about two years they elected to move to Toronto. Lori of course had never lived in the city and Vicki could hardly remember it.

However, they researched the GTA for a new business opportunity and the best location in the area. They decided on pet grooming as this did not require the 24/7 attention that the pet boarding business did. They concluded that Markham had the right demographics and formed a new company called Shear Purrfection.

They moved back with us for a few months while they set up the business. They then bought a house together. Shear Purrfection opened on June 22, 1988 on a very rainy evening. I remember speaking at the opening noting it was appropriate that it was raining cats and dogs. The opening was attended by the Mayor of Markham, Carole Bell.

The business originally had planned to provide an 'in the home' service and it had acquired a new van for that purpose. I also noted that this was one of the few businesses that would be successful if it 'went to the dogs'.

It was a success and still operates now in conjunction with Unionville Veterinary Hospital on Highway 7.

Lori in the meantime had met a young fireman whose family was from Germany in the region near Hamburg. Jurgen Wittemeier's father, Wilhelm, had been in the ship building business in Germany and his mother, Eva, was originally from Hungary.

Perhaps it was the 'German genes' for once again a member of the Fierheller family had married into another German family.

Lori and Jurgen were married on April 3rd, 1993. The wedding was attended by a number of the Wittemeier relatives from Germany.

They now live in Markham.

Vicki and Lori sold their jointly owned house and Vicki bought a house of her own in Ajax. Lori and Jurgen now have two children, Christian, who was born on the 17th of January 1995 and Caitlin who was born on the 14th of March 1997.

Glenna, Travel and the Dogs:

Glenna and I have traveled a great deal including such bizarre adventures as cruising the Southern Indian Ocean, riding jet boats in New Zealand, taking hot air balloon trips in France and Africa, white water rafting north of the Arctic Circle in Sweden, racing Tuk Tuks in Thailand, bouncing on camels in Morocco and even riding an elephant in India.

Glenna, very wisely, did not participate in all these activities. She was quite willing to let me do what I wanted as long as my insurance was paid up. On one trip to Australia it turned out to be far too hot, e.g. over 40 degrees centigrade, to climb Ayer's Rock. I insisted on at least seeing the top and talked a couple of my fellow travelers into chartering a small single engined aircraft to fly over the Rock. We took off at about 1:00 p.m. Glenna suggested we wave as we flew over the pool by the hotel as that was where she was planning to be. I had not counted on either the extreme heat or the thermals that proceeded to toss the plane all over the sky. I think the only pictures I got were the back of the pilot's head and the underside of the wing. As usual Glenna had chosen the best way to spend the afternoon.

I always thought of myself as a romantic. During an around the world private jet trip, Glenna and I stopped at the Taj Mahal. While in New Delhi, I had seen a star sapphire ring which I bought with the intention of giving it to Glenna in that romantic setting. Without taking away from the obvious beauty of the building we soon discovered that it was anything but the picture of solitude one normally sees when Princess Diana was sitting gazing at the structure. We arrived on a holiday with tens of thousands of visitors, beggars and unbelievably pushy peddlers. After a short visit, I put the ring back in my pocket and gave it to Glenna on the airplane to Myanmar. So much for romance.

Whether rings or anything else, I have never been a good shopper. I always assume that if something is more expensive, it must be better. I was buying an overcoat at Harry Rosen's with Harry as the salesman. He advised me that he had just bought three very exclusive scarves in the UK and one would go with the coat I had just bought. I asked him how many he had left. "Three" he admitted.

I agreed to take one without asking the price. I should have known better. It was $350.00. As I recall I never actually wore it because I was afraid someone would steal it if I checked it anywhere. Glenna usually does not let me shop on my own for this reason.

During our many travels, Glenna and I spent some time in China. As I was always looking for something new to do that was not in the tourist brochures, I came across the suggestion that foreigners could visit a small arms testing facility run by the Peoples' Republican Army. Several fellow members of the Young Presidents' Organization agreed to join in the venture. They however were all defrocked Marines who seemed very knowledgeable about guns.

Glenna as usual went along although with some reluctance. I was becoming a little apprehensive as I assumed I would be outclassed entirely by this group.

When we arrived at the facility outside Beijing, we discovered you could buy as many rounds as you wished of anything from hand guns to automatic weapons to bazookas. Our hosts spoke very little English but we soon got the idea. Their motivation was of course to sell arms to foreigners.

To the horror of my Marine friends, I cleaned up in the target practice, leaning on my Dominion marksman experience. Our Chinese hosts were so delighted with my performance that I am now an honorary member of the North Chinese Rifle Association, complete with a red silk bound certificate. I still have the targets!

I was always very sensitive about separating the costs of business trips from any personal travel but at one time at the invitation of Ericsson, I did take Glenna along on a business trip to Sweden. The Ericsson executive group had arranged for a trip to Kiruna north of the Arctic Circle to visit some Lapp facilities.

One part of the trip involved a helicopter ride over a glacier. Glenna had never been fond of helicopters and this one was piloted by a young Swedish hot-rodder. He would swoop down to see some reindeer or whatever looked interesting. At the end of the trip I asked Glenna how she had enjoyed the views of the glacier. "What glacier" she asked. I do not think she had opened her eyes from the moment we took off.

However, as could be expected on one of our trips which was supposed to be a wine tour, she talked me into going to the world's largest dog show at Crufts in London. In those days the show was held at Earl's Court. There she saw and fell in love with Chinese Cresteds.

The CC's as they are sometimes known are unusual animals. They weigh about 10 pounds, have a mane something like a horse, a plume on their tail and socks on their feet. The rest is bare skin. There is a full coated version called a Powder Puff that is equally nice as a pet but not as spectacular.

The breed was not recognized by the Canadian Kennel Club. Glenna imported what turned out to be the matriarch of the breed in Canada from a well known breeder in New York. As a largely naked dog she was appropriately named Lady Godiva.

Glenna eventually got the breed recognized by the CKC and Lady Godiva went on to become the first Canadian Champion Chinese Crested. She lived to be 16 and was the Grand Dame of the Chinese Crested part of Four Halls Kennels.

While Glenna bred her Maltese or Chinese Cresteds, I felt I needed a dog of my own – a real dog as I called it. I selected a Saluki which I could walk around the block (having no intention of walking a long coated Maltese with pink bows in its hair). We have now had five generations of Saluki.

I recently read an article indicating that not only do people tend to look like their animals but they actually are like their animals. At first I scoffed at the idea until I realized that the Saluki is stubborn, aristocratic, fussy about its food and is generally elitist. Perhaps there is more to this than I thought!

Glenna and I have recently fallen in love with cats, more specifically Devon Rex. The one we have at the moment is typical of the breed with a rippled or wrinkled texture to its coat. Its name, naturally enough is Ripple Van Wrinkle.

The cat is very affectionate and tends to use us as warm blooded furniture like all cats.

One day while changing clothes at Rosedale Golf Club in Toronto, a friend noticed some red marks on my back. "Nail scratches" I explained. He said nothing but as I was 65 he was obviously impressed.

In fact, this was the truth, but not the whole truth. One of our Devon Rex named Waverly (his real name being Short Wave as a concession to my communications career and his wavy coat) loved to climb around my neck while we were eating. He would then 'vulture' my food by stretching down my arm while gripping my back with his hind feet.

I, however, saw no need to elaborate.

Vicki took over the Maltese portion of the Kennel and continues to show and breed very successfully. In total, Four Halls has bred or owned 63 Canadian Champions, 25 American Champions and has won 31 Best in Shows.

So, if this is where we are, is there any wisdom for the future?

At the Show of Shows in Ottawa.

III. The Way We Might Be

Why Are We the Way We Are?

For a family history I have spent too much time on my career but then understandably that is what I know best.

I began by suggesting that knowing one's ancestors should help to explain why we are the way we are. Now I am not so sure.

Certainly some characteristics are recognizable:

From my father I can recognize in me some things that he passed on such as his sense of humour, his generosity, his loyalty to family and friends and his slight shyness.

From my mother I inherited the belief that elitism is not all bad.

From both I had in my upbringing a sense that the world was only going to be as good as you make it, whether what you do is grand or insignificant. The approach can be summarized as:

> If not me, then who?
> If not now, then when?

But many of the things that have occupied my life do not seem to have any particular family precedence. There is no background for my love of classical music, science, history, wine or fine dining. Only the two Parsons spinsters, Flo and Vera, seemed to share my love of travel.

I suppose my interest in Science might have come through Uncle Stanley and perhaps the entrepreneurial approach may have come through my two grandfathers.

The link however is tenuous. Further, none of these interests of mine are major interests of Vicki and Lori. I am beginning to think that I am something of a mutant – genetically modified George! Both daughters of course love animals and it is easy to see which side of the family that came from.

So if I am not passing these things on, is there any legacy I can leave them other than a financial one?

Perhaps there really is an unspoken Fierheller philosophy – so simple and so unoriginal as to verge on being trite.

The Golden Rule:

Some years ago, John Graham asked if I would speak to a group pursuing Ultimate Reality and Meaning. As I always seemed to be giving talks or writing papers on a wide variety of topics, I agreed, admitting that I really had no idea what Ultimate Reality was.

To my horror I saw that many in the proposed audience had S.J. after their names. "John", I lamented, "these guys are Jesuits – they will tear me apart!"

However, I would do anything for John and agreed to proceed. The group went so far as to put an ad in various daily papers inviting any and all to attend my lecture. It even included my picture. My friends started to call me the Maharishi.

Whatever I would say was clearly not going to come from any established religion being as I have noted a fervent Agnostic.

'The Maharishi'.

The first part of my talk was aimed at pointing out that no single religion could possibly be the road to Ultimate Reality. The arguments would be familiar to anyone who had read Bertrand Russell's "Why I Am Not a Christian".

However, I concluded that there was a very simple approach to life that had worked for me and further was reflected in nearly every major religion or philosophy. A list is on page 97.

Very simply it is to:

> "Do unto others as you would have them do unto you."

This is the famous Golden Rule, a rule more often violated than practiced. It is over quoted but seldom over used.

I pointed out that it works for a person of any faith or no faith.

I was once asked how one could make reasonable decisions in unreasonable circumstances. After some reflection, I realized that I had developed a technique that seemed to work for me. I always look at any decision whether large or trivial from the point of view of an observer perhaps ten years in the future. Or one hundred years. Or one thousand years. My point was simply that if those in the future could look back on the decision I was about to make and conclude that the decision was the best that could be made under the circumstances and was made for the right reasons on the right basis, then the decision had stood the test of time and was almost certainly the right one to make.

Another way of looking at it is that regardless of one's religion or philosophy, most of us will have an in-bred feeling about whether or not we are doing the right thing. Simply stated "if it does not feel right, don't do it".

This does not guarantee you will always make the right decision but it does ensure that you can sleep well at nights knowing that you had tried to make each decision based on sound principles including the above mentioned Golden Rule.

The Rev. Terry Finlay, who chaired the meeting and is now the Anglican Archbishop of Toronto, commented that he actually agreed with my conclusion, just not how I got there.

Perhaps this is what has really driven the Fierheller family over the years. It is hardly exclusive to us and would not be really useful if it were.

Both daughters already practice it.

Perhaps it is the only important legacy we can help to pass on.

THE VIERHELLERS IN HESSE

Peter Vierheller
d. after 1635
(Wife and 3 of his children – all died in 1635 of the plague)
Hartmannshain

Claus Vierheller
b. 16 July 1620, Herchenhain Parish
(6 children born in Breungeshain
- Godparent – Claus Junker)

Gorg Vierheller
b. 10 July 1653, Breungeshain
m. Catherina Zinnel
27 May 1681, in Breungeshain
(7 children)

Hans Georg Vierheller
b. 22 October 1694
m. Catharina 18 July 1724
(b. 1700 d. 1 September 1762)
d. 12 January 1774 in Breungeshain
(5 children)

Johann Heinrich Vierheller
b. 27 June 1740
m. 29 November 1763, in Breungeshain
Anna Margaretha Zinnel
(3 children)
No record of his death in Germany
Likely died 27 November 1827 in Markham

Johannes Vierheller
b. 1 September 1764, in Breungeshain
Godparent Johannes Zinnel, blacksmith, brother of his mother
No record of his death in Germany

A Family Philosophy

The Golden Rule As Seen By Others:

Buddhism: "Hurt not others with that which pains yourself." *Udanavarga,* 5, 18.

Christianity: "All things whatsoever ye would that men do to you, do ye even so to them, for this is the law and the prophets." *Bible, St. Matthew* 7, 12.

Confucianism: "Do not unto others what you would not they should do unto you." *Analects* 15, 23.

Hebraism: "What is hurtful to yourself do not to your fellow man. That is the whole of the Torah and the remainder is but commentary." *Talmud.*

Hinduism: "This is the sum of duty: do naught to others which, if done to thee, would cause thee pain." *Mahabharata,* 5, 1517.

Islam: "No one of you is a believer until he loves for his brother what he loves for himself." *Traditions.*

Taoism: "Regard your neighbour's gain as your own gain, and regard your neighbour's loss as your own loss." *T'ai Shang Kan Ying P'ien.*

Zoroastrianism: "That nature only is good when it shall not do unto another whatever is not good for its own self." *Dadistan-i-dinik* 94, 5.

FURTHER READING

Fierheller:

"Markham 1798-1900" – Isabel Champion
Markham Historical Society 1979

"Markham Remembered" – Mary B. Champion
Markham Historical Society 1988

"The Lutherans in North America" – E. Clifford Nelson
Fortress Press – Philadelphia 1980

"The German Americans" – Anne Galicich
Chelsea House Publishers 1989

"German Pioneers of Toronto and Markham Township"
Historical Society of Mecklenburg Upper Canada Inc. 1987

"Canadians in Russia 1918-19" - Roy McLaren
Macmillan of Canada 1976

Parsons

"What's My Line" – W. W. Foot 1978
Published Privately

Hatheway/Bauld:

"Sketches & Traditions of the Northwest Arm" – John W. Regan
Hounslow Press 1978

"The Old Burying Ground, Fredericton, N.B." Vol. 1 – Isabel Louise Hill
Fredericton Heritage Trust 1981

"History of Dartmouth / District Families and Halifax Harbour" – Douglas W. Trider
Ken Mac Print Ltd. 2001

"Early Marriage Records of New Brunswick to 1839" – B. Wood-Holt
Holland House Inc. 1986

"Early Loyalist Saint John" – D. S. Bell 1783-1786
New Ireland Press 1983

Grant:

"The Clan Grant" – I. F. Grant
Johnston & Baron – Edinburgh 1969

Personal:

"Ultimate Reality & Meaning"
Journal of Interdisciplinary Studies in the Philosophy of Understanding
Vol. 16, No's 3 & 4 – 1993

"An Introduction to the Geo-information System of the Canada Land Inventory"
R. F. Tomlinson 1967

"The SDL Story" – G. A. Fierheller 1988

Snippets from the life of George...

Practising the kickoff for the 1981 United Way Campaign in Vancouver.
Lou Passaglia holding the ball.

The launch of cellular service in Canada. Mayor Art Eggleton, Ted Rogers and George. Toronto, July 1, 1985.

The famous Cod Kissing ceremony in Newfoundland.

Winner of the Worst Tie contest for the United Way at Cantel.

Launch of a new wireless product in Ottawa. Michael Binder, ADM Industry Canada; Johnathan Frakes, Commander Wil Riker, Starship Enterprise; and George.

George with Prime Minister Brian Mulroney.

George piloting an airship over Toronto.

Hot air ballooning over the Serengeti.

George with Dr. Anne Golden and friend.

Jet boat thrills in New Zealand.

After the jet boat ride in New Zealand.

George, as the proud sponsor of a Lemon 'Moose', a Toronto fundraiser.

Glass blowing in Sweden.

SDL Systemcentre staff.

Sharing a laugh with Premier Bill Bennett, Vancouver, 1983.

Origin of the Canadian Fierheller family, Buttonville Cemetery, Ontario.

Jurgen and Lori Wittemeier.

Caitlin and Christian Wittemeier.

Vicki with a Chinese Crested.

Tristan and Kuhbah, two of the Salukis.

One of our Maltese.

The Lindop Family: John, Michael, Audrey, Peter and David.

Opening Cantel services with Mayor Elsie Wayne of Saint John, N.B.

Signing the final beam at the topping off of the Royal Ontario Museum 'Crystal'.

With David Johnston, President, and Ken McLaughlin, University of Waterloo examining part of the SDL art collection.

Guarding the University of Toronto Chancellor at a degree granting ceremony for Michael Wilson.